THE BOOK OF TIMES

THE BOOK OF TIMES

From Seconds to Centuries, a Compendium of Measures

LESLEY ALDERMAN

WILLIAM MORROW

An Imprint of HarperCollins*Publishers*

HarperCollins books may be purchased for educational, business, or sales promotional use. For information please write: Special Markets Department, Harper-Collins Publishers, 10 East 53rd Street, New York, NY 10022.

FIRST EDITION

Designed by Diahann Sturge

Library of Congress Cataloging-in-Publication Data has been applied for.

ISBN 978-0-06-207418-8

13 14 15 16 17 ov/rrd 10 9 8 7 6 5 4 3 2 1

For Ann, whose time was too short
And for Steve, who makes all the time better

Contents

THE BOOK OF TIMES

Introduction

My mother was always late. I vividly remember being the last kid picked up from birthday parties and waiting anxiously at airports for her yellow station wagon to appear around the corner. Running late *with* her was no less stressful. Speeding through traffic in a futile effort to make up for her late start, my father would be fuming behind the wheel, and my mother would be doing some last-minute task like mending the hem on her dress. Those minutes seemed like hours.

Time is subjective. Minutes spent waiting multiply; mundane tasks drag on. I recently timed how long it takes me to empty the dishwasher: 5 minutes. I had predicted it would take 15. Yet, blissful experiences, when we're in the flow, seem to speed by. As Einstein observed: "When a man sits with a pretty girl for an hour, it seems like a minute. But let him sit on a hot stove for a minute and it's longer than any hour. That's relativity."

My mother liked to fill time to the brim. It made her chronically late, but also incredibly industrious. It wasn't until I was an adult and met people from other countries and cultures that I realized not everyone operated on the edge of time. Some people think that less is more and allow ample time to get everywhere.

Time is confounding: It stands still. It flies. And it goes in cycles. There is no time like the present but tomorrow is another day. Ever since Einstein, we know that time (along with everything else) is relative. Poets wrestle with it, scientists measure it, and businesspeople try to lasso it into submission. And we all want to be reassured—don't we?—that we are not wasting the finite hours we have before us.

Americans generally feel time pressed. More than 1 out of 4 working adults say they don't have enough time to get done what they need to do, which makes them feel less satisfied with their lives and more stressed out. Meanwhile, the percentage of Americans who believe that science is making life change too quickly has been rising for the past five years.

The faster life becomes, the more we try to catch up with it. We want smarter phones, higher-performing computers, and faster food. We pride ourselves on multitasking and productivity. But then we flock to yoga and meditation classes, spin records the old-fashioned way, and cook slow food in an attempt to connect with the present moment. It's a constant struggle between our desire for instant gratification and our need for peace and reflection.

What to Expect?

This is a book about how long things take, how long things last, and how much time we devote to a variety of activities. Why? Because objectively measured time is so surprising. How long does it take to build a bridge, write a symphony, or make a million dollars? What's the optimal amount of time for making love? What's the real time? And how are we spending the hours of each day?

The Book of Times is meant to amuse, inform, and provoke—think of it as a quirky collection of facts with a slightly political edge. You may find yourself saying, "Wow, who knew?" or "No way, this can't be possible!" You may feel angered or even annoyed by facts or trends (that's the political part), and I hope you will find yourself questioning the way in which you spend your precious time each day.

Why a Book on Measured Time?

Perhaps because of my mother's peculiar attitude toward time, I have always been acutely aware of it. As a child I would write my schedule for the evening on a black chalkboard in my room: 6:00 P.M. homework, 7:00 P.M. call Karen, 7:30 P.M. *Gilligan's Island*. You get the idea.

As I got older, I used time in an attempt to make order out of chaos. I timed how long it took me to get to the nearest subway stop, to read a page of a book, and to fold a pile of laundry. I even timed my romances: after three months of dating, if I was not in love, then it was time to move on.

The Book of Times sprang from my personal fascination with measured time and then morphed into something broader—a collection of measurements of all sorts of things, from love affairs to school days, from the time we spend waiting to the time we spend learning. How many things can time measure? How many measures of time can be found? And what can those measures reveal?

Lucky for me, Americans love to gather data. The government measures the time we spend on all sorts of activities, as do polls, surveys, scientific studies, and academic research.

But just like time, data are relative. This book is chock-full of measurements, but not all measurements are created equal and not all measurements will stand the test of time. So I invite you to read on with an open, incisive, and questioning mind. And if you have gripes, suggestions, or corrections, please send me an e-mail: Lesley.Alderman@gmail.com.

Helpful Hints

The Book of Times is laid out in twelve chapters, like the twelve hours on the clock. Each chapter covers a single subject—love, family, energy, and so on—and contains dozens of timings. You'll also find suggestions on how to become more mindful of time through Time Exercises, quotes from great thinkers, and quizzes.

Before you begin reading, here's a quick refresher on some basic statistical terms and strategies that appear throughout the book. (If you have a math mind, skip ahead.)

When statisticians want to find the center of a set of data (commonly known as the average), they may measure the mean, median, or mode—and sometimes all three.

- **Mean** is interchangeable with the term **average**. To find the mean or the average of a data set, researchers tally up all the data and then divide by the total number of data points.

- **Median** is the middle value of a list of numbers that are in a sequential order. Mean scores can be affected by outlier numbers at the edge of the data set, so when there is a set of numbers with extreme values, it's often better to use the median to find the center of the set.

- The **mode** is the value that occurs most often.

When researchers measure how people spend their time, they employ a few different strategies to get precise data. Let's say statisticians want to find out how much time Americans spend reading books: they could tally up the minutes every American admits to reading each day and provide a straight average, or they could tally up just the subset of Americans who actually read each day. The first average would include readers and non-readers and be on the low side, while the second average would include just readers and provide a higher number. See the difference?

Statistics are tricky, and researchers often present the data in a light that supports their specific agenda or point of view. When I use data that might be skewed—say by the size of the sample, the date, or the authors' point of view—I'll say so. "We are all mediators, translators," wrote the philosopher Jacques Derrida.[1] I've tried to get past perceptions and represent the best reality possible. Here, then, is my particular and carefully curated collection of timings.

Daily Life
Schedules, Happiest Times & Waiting

"How we spend our days, is, of course, how we spend our lives," the writer Annie Dillard observed. Broadly speaking, a day in the life of a typical American looks something like this: one-third spent sleeping, one-third spent working, and the remaining one-third spent on doing chores, making meals, eating meals, caring for kids and parents and pets, watching TV (a lot), socializing, and shopping. How people choose to spend those 8 hours of discretionary time varies tremendously depending on their age, gender, marital status, and more. The good news: leisure time has increased over the past four decades. The bad news: Americans don't feel like it has.

As you read this chapter, consider this: Is the way in which you spend your time the way you actually *want* to spend your time?

Time Spent . . . The Big Picture

Let's start with a look at how adult Americans spend their days. The following table shows the number of hours that people who actually participate in a specific activity spend doing the activity. The category "Volunteering," for example, only includes those people who do volunteer work, which makes the numbers more robust than a straight average of everyone in the population.

The numbers, collected by the government's American Time Use Survey[1], reflect time spent on one activity exclusively. If you are watching TV *and* texting your friends, the time gets allocated to "watching TV" and not to e-mail as well. This way of collecting data is not very twenty-first century!

Activity	Weekday	Weekend
Sleeping	8.45 hours	9.36 hours
Working	8.39	5.88
Education (classes and homework)	6.17	3.45
Watching TV	3.31	3.98
Volunteering	2.28	2.86
Taking care of household members	2.10	1.99

Lawn and garden care	1.98	2.36
Other	1.80	1.96
Taking care of people outside the household	1.76	1.80
Socializing & communicating	1.66	2.56
Purchasing stuff	1.62	1.87
Housework	1.60	1.96
Participating in sports, exercise & recreation	1.47	2.06
Religious activities	1.30	2.04
Eating & drinking	1.23	1.43
Food prep & clean up	1.00	1.15
Personal care	.82	.77
Telephone, e-mail, and mail	.71	.80
Household management	.71	.88

Leisure Time

How much do we really have?

How many minutes do Americans have that are free of chores, work, and other responsibilities, such as child care? "Not enough!" is the resounding response of most Americans. And yet one notable study published in 2006 by economists Mark Aguiar and Erik Hurst concluded that leisure time has increased by 6 to 8 hours a week for men, and 4 to 8 hours for women over the past five decades.[2] Why the increase? The authors found that men now put in fewer hours of paid work, and women spend less time on housework.

Ironically, over approximately that same five-decade time span, satisfaction with leisure time fell. In 1963, 76 percent of Americans were satisfied with the amount of leisure time they had; in 2010, just 65 percent were, according to a *USA Today/Gallup* poll.[3] Why such dissatisfaction? While we might have more hours of leisure, we have less hours of *pure leisure*—that is time spent doing one enjoyable activity without interruptions (such as text messages) or an added activity (like watching your child). What's more our leisure time is more fragmented, which makes it feel less satisfying.[4]

Favorite leisure activities

Americans' favorite pastimes are watching TV, spending time with the family and kids, and exercise (in that order), according to a 2008 Harris poll.[5] If the poll were conducted in 2013, I suspect "using the iPad" might be on this list.

Not all leisure time is equally pleasurable

Leisure time is the hours left after all obligations are taken care of. You can do something passive, like watch TV, or something active, like take a walk or read a book. Both types of leisure are rewarding, but one may make you happier than the other, according to research conducted at Princeton University. Americans rate *active* leisure activities (e.g., exercise, bird watching) as more enjoyable than *passive* ones (e.g., watching TV), and yet they spend more time on passive activities. Let's face it: it's simpler to sit down in front of a screen than to lace up your boots and go for a walk. Active leisure may be more rewarding, but passive leisure is more accessible—and after a long day of responsibility, accessibility may trump all else. But over the long term, making the effort to indulge your hobbies or passions may be more satisfying than watching *American Idol*.

 Time Trivia: Which country spends the most time relaxing and doing nothing?

Italy and Spain tie for first place! A survey by the Centre for Time Use Research looked at time usage in 22 countries and found that Italians and Spaniards spend 1 hour and 17 minutes a day relaxing and doing nothing (let's call it R&DN), followed by Turkey (1 hour, 14 minutes), Slovenia (1 hour, 7 minutes), and the United States (1 hour, 6 minutes). At the other end of the spectrum were the Belgians, who spend 34 minutes a day R&DN, followed by France (39 minutes) and Norway (46 minutes). Before you rush to judgment, bear in mind that societies define leisure differently. For instance, though the French appear to spend relatively little time R&DN, they spend comparatively more time in restaurants. Perhaps the French feel so relaxed during mealtime that they don't require downtime. And while the Turks take a lot of time to chill, they spend very little time socializing with others outside their homes.[6]

Doing nothing can be a waste of time, or it can be an art form.
—Leo Babauta, creator of the website ZenHabits.net

Who spends the most time watching TV?

Americans *love* to watch TV, or maybe they don't *love* it but they spend a lot of time doing it: about 3 hours a day, on average. (See the Media chapter for more details on Americans' TV habits.)

 Time Exercise 1: Track Your Computer Time

Wondering where your computer time goes? Try RescueTime .com, a free online productivity tool. You tell the site which activities you consider productive and which nonproductive and the program keeps track of how and where you allocate your time. The program will tell you how many hours you work each day, what percentage of those hours are spent in activities you deem productive, and which day of the week is your most prolific. If you upgrade to the fee service, the program will even block "distracting" sites for specific amounts of time. Of course, you could just watch the clock and police yourself—but what fun is that?

Shopping Time

Is shopping a leisure activity or a chore? The answer depends on who are you and why you shop. Whether you love it or hate it, most Americans clock a lot of time shopping. We spend 1 hour and 40 minutes each month shopping for clothes in stores and slightly more time, 1 hour and 43 minutes, shopping for new duds online, according to Cotton Incorporated's Lifestyle Monitor Survey.

We also spend a lot of time at the mall. Over the past decade Americans have made the same number of monthly mall visits (3), but spend more time per visit, 89 minutes in 2010 versus 78 in 2000. Overall, shoppers spend nearly 4.5 hours at the mall each month, according to the International Council of Shopping Centers.

Primping Time

How much time do men and women spend getting themselves gussied up for the day ahead? One study compared the grooming habits of American workers and found that, on average, minority women spend 55 minutes a day on grooming, while white women spend just 47 minutes. Minority men spend 37 minutes on preparing themselves for work, while white men spend just 32 minutes, according to the research, which was conducted by two economists at Elon University in North Carolina.[7] (Sneak preview: In Chapter 10, you'll learn the answer to this question, *How does grooming time relate to weekly earnings?*)

Time Trivia: Chewing Time, Chimps vs. Humans

Because humans cook most of their food, they only spend 1 hour a day chewing it (lucky us). Chimpanzees, on the other hand, eat raw food from the wild and thus spend about 6 hours a day chewing, according to *Catching Fire: How Cooking Made Us Human.*[8]

Women's Time

Here are some quirky findings about how women spend their time. You'll find more serious (and less sexist) findings in Chapters 3 and 6.

How much time do women spend getting ready for work during the week?

According to a British survey, women in the United Kingdom devote, on average, 1 hour and 16 minutes to primping and dressing on Monday, and a mere 19 minutes on Friday. Women in the United Kingdom might be different from women in the United States, but the study has a universal ring of truth. On Monday, women may put effort into their looks to get their week off to a good start, but by Friday, they may be worn out and thinking, *Eh, who cares; I'll just wear jeans.*[9]

Primping activity	Monday	Tuesday	. . . Friday
Hair	23 minutes	12	6
Makeup	18	9	2
Getting dressed	16	9	1
Showering	19	10	10
Total time	**76**	**40**	**19**

How much time do women spend chatting and gossiping?

Five hours a day! According to one survey, women spend 5 hours a day gossiping and chitchatting with friends. The survey, which was commissioned by winemaker FirstCape, also found that women spend an additional 24 minutes a day discussing their weight, diets, and dress size.[10]

Teen Time

Teens spend a lot of time sleeping, playing games, and watching TV, but they also log some serious study hours. High school students spend one-fourth of their day in school or dealing with homework. Even so, most parents wish their teens spent more time reading and pursuing healthy recreational activities and less time playing video games, watching TV, and getting into trouble.[11] (For more on how much time teens spend on media-related activities, flip to Chapter 11.)

How do high school kids spend their days?

Like their adult parents, teens watch a lot of TV. This table shows how much time high schoolers who engaged in select activities spent doing the activity. (Kids who don't work, for example, are not included in the "Work" line.) Check out the gender differences: Guys spend more time playing games, for example, and girls spend more time relaxing and thinking.[12]

Activity	Guys	Gals
Sleeping	9.19 hours	9.17 hours
School and homework	6.65	6.40
Work	3.23	3.93
Playing games	2.86	1.75
Watching TV	2.75	2.91
Sports/Exercise	2.45	2.03
Socializing & communicating	1.25	1.48
Relaxing & thinking	1.04	1.41
Housework	0.93	1.00
Caring for household members	NA	0.96
Eating and drinking	0.86	0.95
Reading: personal	0.75	1.04
Shopping	0.65	1.08

High schoolers snooze most on Sunday

Tired teens sleep for 10.7 hours on Sunday, more than any other day of the week. They sleep just 8.4 hours on Fridays (also known as party night number one!), the fewest nightly hours during the week.

The future is something which everyone reaches
at the rate of 60 minutes an hour,
whatever he does, whoever he is.
—C. S. Lewis, novelist and scholar from Belfast, Ireland

Pet Time

How much time do pet lovers spend with their critters?

Roughly 40 minutes a day; that number has been fairly consistent over the past 7 years. In 2003, men spent 38 minutes a day grooming, feeding, and walking their furry friends, while women spent 37 minutes. Fast-forward forward to 2011 and the numbers change just slightly: men devote 42 minutes to their critters, women 36 minutes.[13] Pet spending, though, is escalating. The average household devotes more dollars to pets than retirement savings or alcoholic beverages.[14]

Religious Time

How much time do Americans spend on religious activities?

Well, 35 percent of Americans say they attend a church, mosque, or synagogue at least once a week; 25 percent say they seldom attend; and 20 percent never attend, according to a Gallup poll. Attendance varies by affiliation and group. Conservatives, blacks, Republicans, and those over 65, for instance, attend church frequently; by comparison, liberals, Asians, young adults, and singles attend religious services infrequently.

The Gallup pollsters also found that Americans who identify themselves as *very religious* have higher personal well-being than those who ID themselves as moderately religious or not religious at all. Food for thought.[15]

How often do Americans pray?

More than half of Americans, 58 percent, pray daily. Folks in southern states are the most likely to say they pray, while those in the Northeast are the least likely. Mississippians pray the most (77% of the state's residents pray daily) and Mainers pray the least (just 40% of residents pray each day). Young people pray less than older adults; just 48 percent of Americans between the ages of 18 and 29 say they pray daily.[16]

How often do Americans ask for God's help?

Quite a bit: 23 percent of Americans ask for God's help "many times" a day; another 24 percent ask for help every day. A stoic or atheistic 14 percent of Americans never, or almost never, ask for his help, according to the General Social Survey.[17]

Volunteer Time

How much time do Americans devote to volunteer work?

Volunteer time depends, in part, on age and education. Older folks tend to volunteer more than younger folks and, according to one study, educated people are more likely to volunteer, but they may spend fewer hours per day on their good deeds than those with less education.[18]

	Percent who volunteer	Average daily hours spent on volunteer work
Less than a high school diploma	2.9	2.7
High school diploma	5.4	2.2
Some college or associate degree	6.9	2.2
Bachelor's degree or higher	10.8	2.0

Three o'clock is always too late or too early
for anything you want to do.
—Jean-Paul Sartre, French philosopher

Culture Time

How much time do Americans spend at museums and libraries?

Not much. According to the General Social Survey, Americans are not hanging out regularly at cultural institutions.[19]

How many times did you visit . . . last year?	Percent who said not at all	Percent who said once
. . . a public library	36	9
. . . a zoo	51	27
.. an art museum	67	15
. . . a natural history museum	74	18
. . . a science museum	76	17

How much time do Americans spend attending arts performances?

The portion of American who attend performances is small, but those who do spend a considerable amount of time at them. For instance, on an average day, just .6 percent of American adults go to a performance, but those who do spend an average of 2.6 hours enjoying it.

Nearly half of Americans go to at least one arts activity in a given year. Attendance increases with income: just 27 percent of those with incomes below $10,000/year go to performances, compared to 78 percent of those with incomes over $150,000. Here's where people spend their time.[20]

Arts Activity	Percent of Americans who attend in a given year
All arts attendance activity	49
Art museums	23
Musicals	17
Classical music	9
Non-musical plays	9
Jazz	8
Ballet	3
Opera	2

How much time do we spend at the library?

Yes, libraries still exist, but barely. According to a *60 Minutes/Vanity Fair* poll, 35 percent of Americans never go to libraries; 31 percent go once or twice a year; 18 percent go once a month; and 15 percent go once a week. Libraries are still a great place for children as well as for adults who want peace and quiet or a place to study or write (I wrote a portion of *The Book of Times* at my local Brooklyn library), but they are not the hub of family life that they once were.

Waiting Time

One urban legend holds that we spend 3 years of our life waiting. It's a great stat, but likely untrue. It may *feel* like we spend years of our life on lines because we perceive waiting times as exponentially longer than they really are.

Time spent waiting . . . on store checkout lines

Who spends the most time waiting on store lines? New Yorkers. According to a 25-city survey, New Yorkers spend an average of 6 minutes and 51 seconds waiting on individual store lines. Miami comes in second at 6 minutes and 44 seconds. Who spends the least time? Clevelanders—they spend just 4 minutes and 33 seconds on line.[21] Which lines do people hate the most? Grocery store check out lines, according to an Ipsos survey from 2006. The same survey also found that half of consumers have refused to return to stores that had long wait times.

. . . to see doctors?

The average wait time is 24 minutes. But in urban areas and among certain specialties, the waits can be much longer. Neurosurgeons have the longest wait times—30 minutes, on average, according to Press Ganey's *2010 Medical Pulse Report*. Why? "All patients have questions for their doctor, but not surprisingly surgical patients have more questions about the procedure, process and expected outcomes," the authors of the report wrote.[22]

. . . in the emergency room?

The average wait time was a staggering 4 hours and 7 minutes in 2009. Patients in Utah waited the longest, over 8 hours, while those in Iowa faced the shortest waits—just shy of 3 hours.[23]

Hate waiting? Don't move to Russia

It may not be a communist country anymore but Russians still spend a lot of time on lines. The Mystery Shopping Providers Association sent its spies to wait on lines in stores, banks, post offices, and drugstores in 24 European countries and found that Russian customers spent the most time waiting on queues, followed by the Italians and Bulgarians. The Swedes got off easy, just 2.2 minutes per line.

Country	Average wait time per line
Russia	27.1 minutes
Italy	14.4
Bulgaria	10.9
Turkey	10.4
Croatia	9.6
Romania	7.9

Hungary	7.2
Poland	6.6
France	6.4
Greece	6.0
Portugal	5.8
The Netherlands	4.7
Latvia	4.6
Ireland	4.3
Macedonia	4.1
Estonia	4
Lithuania	4
Spain	3.3
UK	3.3
Denmark	3.1
Sweden	2.2

Book Excerpt: Waiting etiquette across cultures

"Different cultures have different expectations of the waiting experience. One major difference is whether there should be a line at all. Polite, orderly queues are the rule in some cultures. In others, people try to force themselves to the front, with the noisiest or most forceful folks winning. Travel around the world and you will find the differences are striking: orderly queues of patient people in London; disorderly mobs clamoring for train tickets in Beijing and Casablanca. In much of Asia people will crowd around counters, each person demanding the attention of the service providers. Although many Westerners are appalled, the system works well. A Chinese friend explained that with a typical, orderly (Western) line, people wait for a long time with nothing happening. In the apparent disorder of the Eastern crowd clustered around the service agents, people can get attention almost immediately. Although the agent's attention to their problem is quickly interrupted by the demands of other people, a tiny amount of the transaction gets accomplished. In the end, both systems may take equally long, but in the Asian method there is a continual feeling of progress."

"Cultures can be changed. McDonald's changed queuing behavior in Hong Kong: The social atmosphere in colonial Hong Kong of the 1960s was anything but genteel. Cashing a check, boarding a bus, or buying a train ticket required brute force. When McDonald's opened in 1975, customers crowded around the cash registers, shouting orders and waving money over the heads of people in front of them. McDonald's responded by introducing queue monitors—

young women who channeled customers into orderly lines. Queuing subsequently became a hallmark of Hong Kong's cosmopolitan, middle-class culture. Older residents credit McDonald's for introducing the queue, a critical element in this social transition."

From "The Design of Waits," Chapter 7 of Donald Norman's *Living with Complexity* (Cambridge, MA: MIT Press, 2010). Reprinted with permission of Donald Norman.

What makes wait times fly?

Being treated nicely, for starters. In one study, customers who were verbally greeted when they entered a store recalled that they waited on line for an average of 3.42 minutes, while those who had not been greeted said they had waited 9.27 minutes—nearly three times longer. A similar effect was found when customers were welcomed with a smile.[24] Annoyed customers generally overestimate the time they are kept waiting by 23 to 50 percent.[25] Distractions help make waits fly, which is why many stores play music and have cool things to look at on the check out line such as magazines, knick knacks, and candy.

Happiness and Time

Does more free time lead to more happiness? You might think so, but it's the *quality* of free time that's important, rather than the *quantity*.

More social time = more happiness

The more time people spend with friends, the higher their happiness levels and the lower their stress levels, according to the Gallup-Healthways Happiness-Stress Index. This squares with recent research that has found lonely people tend to have poorer health. Gallup's national survey questioned highly social people and highly unsocial folks. Here's what researchers found:

Of people who spent 6 to 7 hours a day socializing (where *do* they find the time?):

Over one-half reported feeling a lot of enjoyment and happiness; just 5 percent experienced stress and worry.

Of people who were alone all day:

Just one-third said they had experienced a lot of enjoyment and happiness; one-third reported feeling stressed.

Time Trivia: Talk Time

What you choose to spend your time talking about may be related to how happy you are. Serious talkers—those who spend most of their time discussing politics, the economy, religion, health care reform, and education—report being happier than light talkers. A 2010 Harris Interactive poll, found that 39 percent of serious talkers are *very happy,* but only 32 percent of those who chat mostly about sports, fashion, and celebrity gossip are similarly happy. Does talking about serious things make folks happy, or do happy people talk about serious things? The poll can't establish a causative effect, but the results certainly seem counterintuitive.

How much time do we spend with friends?

Hard to say. According to government stats, we spend 2 hours a day "socializing and communicating," but that number does not account for the time we spend with pals at work, at school, or at sporting events. Friend time often overlaps with social time, with school time, and with work time. It's a tricky number to parse.

But here's an interesting stat to consider. Robin Dunbar, an evolutionary anthropologist and author of the book *How Many Friends Does One Person Need?*, says that people spend 40 percent of their limited social time each week on the five most important people they know, and those five people represent just 3 percent of their social world. "Since the time invested in a relationship determines its quality, having more than five best friends is impossible when we interact face to face one person at a time," Dunbar wrote in an opinion piece for *The New York Times*.[26] Humans, Dunbar believes, have the capacity for just 150 friends: "It has been 150 for as long as we have been a species. And it is 150 because our minds lack the capacity to make it any larger," he wrote in his book.[27] Tell that to folks on Facebook. The average FB user has over 200 "friends."

Time formula for happiness?

Yeo Valley, a British dairy company, conducted an ambitious survey of 4,000 Brits to find out which lifestyle habits make people the happiest. The company also asked participants what was the optimal amount of time to spend on daily activities. Based on the results of their research, the pollsters came up with a formula for a happy day. (Note, times add up to more than 24, presumably because some activities can be combined, such as spending time outdoors and playing with children.)

How do these numbers square with your idea of a perfect day?

Activity	Daily hours
Work time	7 hours and 15 minutes
Uninterrupted sleep	6 hours and 15 minutes
Partner time	3 hours and 58 minutes
Outdoor time	2 hours and 49 minutes
Playing with children	2 hours
Watching TV	2 hours
Exercise	23 minutes a day (the survey stated 2 hours and 45 minutes a week)
Commuting to work	20 minutes

Happiest Days, Years, and Decades

The happiest days of the year

The happiest days are those that are on or close to holidays. Here's the list from 2011, ranked in order of *most* happy-making.[28]

1. Christmas Day, December 25 (presents!)

2. Thanksgiving (food!)

3. Easter Sunday (candy!)

4. July 4th: Independence Day (fireworks!)

5. New Year's Day (zzzzz)

6. Day after Christmas (ZZZZ)

Happiest times of the day

A study of Twitter users found an interesting pattern: humans tend to be happy at breakfast time, not so happy at midday, and then happy again near bedtime. The study, which analyzed 509 million tweets from 2.4 million users in 84 countries, found that moods fluctuate in a predictable pattern. On weekdays, positive tweets peak between 6 A.M. and 9 A.M., then decline steadily to a trough between 3 P.M. and 4 P.M. In the late afternoon, positivity begins to rise again, peaking after dinner. On weekends, the pattern is similar but morning happiness shifts later, starting at around 9 A.M., when most people are beginning their day. The study's authors used a text-analysis program that scanned the tweets for words that had positive and negative affects. The moral: we should all be taking a siesta at 3 when life is most trying.[29]

Happiest years of life

There's some debate, and the answer often depends on the age of those you ask. Friends Reunited, a British social networking website, surveyed 2,000 adults and found that 70 percent said they were not truly happy until they hit 33. But the majority of respondents were in their 40s.[30] When AARP surveyed 4,000 adults over age 35 it found a U-shaped happiness curve. Happiness was fairly high in the early 40s, dropped off in the mid-40s, and hit a low in the early 50s. Happiness climbed again, reaching a peak around the mid-60s (retirement age!). If your 30s and 40s were not that stellar, take heart: you may have some great years ahead.

Happiest decade

The 80s and 90s! A *60 Minutes/Vanity Fair* poll asked Americans what was the best decade to live through, from 1930 to 2000. The 1980s got the most votes, followed by the 1990s, and 1950s. The worst? The depressing 1930s.

How Fast Is Your Life?

That depends, in part, on where you live. If you dwell in Boston, your life may feel a lot speedier than if you reside, say, in Salt Lake City. Robert Levine, a social psychologist with a keen interest in time, developed formulas for measuring the pace of life in cities and countries across the globe. In his book *A Geography of Time*, Levine came up with a methodology for measuring the pace of life in 36 American cities and in 31 countries. Levine found that, "People are prone to move faster in places with vital economies, a high degree of industrialization, larger populations, cooler climates, and a cultural orientation toward individualism." (Which explains why New York City feels so speedy.)[31]

The pace of American cities

To determine the pace of life in U.S. cities, Levine looked at a few key measurements: he clocked the walking speeds of pedestrians in the downtown area during business hours, timed how long it took for bank clerks to make change, recorded the response of postal clerks to basic questions, and counted the number of people wearing wristwatches. Here are the results, which—full disclosure— are from 1998; it's possible that paces may have changed since then. Certainly far fewer people wear wristwatches now. It's hard to believe that Boston is faster than New York City, but perhaps it's the influence of all those energetic college kids.

City	Overall Pace	Walking Speed	Bank Speed	Talking Speed	Watches Worn
Boston, MA	1	2	6	6	2
Buffalo, NY	2	5	7	15	4
New York City, NY	3	11	11	28	1
Salt Lake City, UT	4	4	16	12	11
Columbus, OH	5	22	17	1	19
Worcester, MA	6	9	22	6	6
Providence, RI	7	7	9	9	19
Springfield, MA	8	1	15	20	22
Rochester, NY	9	20	2	26	7

Kansas City, MO	10	6	3	15	32
St. Louis, MO	11	15	20	9	15
Houston, TX	12	10	8	21	19
Paterson, NJ	13	17	4	11	31
Bakersfield, CA	14	28	13	5	17
Atlanta, GA	15	3	27	2	36
Detroit, MI	16	21	12	34	2
Youngstown, OH	17	13	18	3	30
Indianapolis, IN	18	18	23	8	24
Chicago, IL	19	12	31	3	27
Philadelphia, PA	20	30	5	22	11
Louisville, KY	21	16	21	29	15
Canton, OH	22	23	14	26	15
Knoxville, TN	23	25	24	30	11
San Francisco, CA	24	19	35	26	5

Chattanooga, TN	25	35	1	32	24
Dallas, TX	26	26	28	15	28
Oxnard, CA	27	30	30	23	7
Nashville, TN	28	8	26	24	33
San Diego, CA	29	27	34	18	9
East Lansing, MI	30	14	33	12	34
Fresno, CA	31	36	25	17	19
Memphis, TN	32	34	10	19	34
San Jose, CA	33	29	29	30	22
Shreveport, LA	34	32	19	33	28
Sacramento, CA	35	33	32	36	26
Los Angeles, CA	36	24	36	35	13

The pace of life across the globe

To measure the pace of life across different countries, Levine took slightly different measurements: he looked at how quickly pedestrians walked in downtown areas; how quickly postal clerks helped customers purchase a stamp; and the accuracy of bank clocks (to measure a city's interest in clock time).

How did the United States end up in sixteenth place, *behind* Italy and France? The USA was represented by New York City and NYC is a very speedy place. But the pace of life may also be translated as the efficiency of life—and life in NYC is not that efficient.

Country	Overall pace of life	Walking Speeds	Postal Times	Clock Accuracy
Switzerland	1	3	2	1
Ireland	2	1	3	11
Germany	3	5	1	8
Japan	4	7	4	6
Italy	5	10	12	2
England	6	4	9	13
Sweden	7	13	5	7

Austria	8	23	8	3
Netherlands	9	2	14	25
Hong Kong	10	14	6	14
France	11	8	18	10
Poland	12	12	15	8
Costa Rica	13	16	10	15
Taiwan	14	18	7	21
Singapore	15	25	11	4
USA	16	6	23	20
Canada	17	11	21	22
S. Korea	18	20	20	16
Hungary	19	19	19	18
Czech Republic	20	21	17	23
Greece	21	14	13	29
Kenya	22	9	30	24
China	23	24	25	12

Bulgaria	24	27	22	17
Romania	25	30	29	5
Jordan	26	28	27	19
Syria	27	29	28	27
El Salvador	28	22	16	31
Brazil	29	31	24	28
Indonesia	30	26	26	30
Mexico	31	17	31	26

Turning

Going too fast for myself I missed
more than I think I can remember

almost everything it seems sometimes
and yet there are chances that come back

that I did not notice where they stood
where I could have reached out and touched them

this morning the black shepherd dog
still young looking up and saying

Are you ready this time

—W. S. Merwin

Love
Romance, Sex & Marriage

"The deepest need of man . . . is to overcome his separateness, to leave the prison of his aloneness," observed the psychologist Erich Fromm in his 1956 book *The Art of Loving*. To be human is to desire love and union. No wonder then, that we spend countless hours thinking about, chasing after, and obsessing over love and all it concomitant parts. Many aspects of love are immeasurable—of course!—but some do have beginnings and ends. Scientists have found, for instance, that the rush of romance lasts for a predictable amount of time. Researchers *almost* agree on the optimal amount of time that people should spend having sex. Suspend judgment and read on.

Romance . . .

Is there such a thing as love at first sight?

Perhaps. When a person sees a potential mate, it takes seconds for the brain to create a chain reaction that registers excitement. In his book *Blink: The Power of Thinking Without Thinking*, Malcolm Gladwell explains that humans have a built-in ability to size people up in an instant; we make snap decisions in just 2 seconds. Love can happen quickly and intuitively. Is that love trustworthy? That's another matter entirely.

How long does love last?

As little as a minute, or as long as a lifetime.

On average, though, the first rush of love seems to last one to two years. Neurobiologists have studied the brain chemistry of the newly love struck and found that certain chemicals are elevated when love is new and that these chemicals dissipate over time. Researchers at the University of Pavia, for instance, found that levels of nerve growth factor (NGF)—a protein that maintains the health of neurons—were higher in people who had reported just falling in love when compared to single people or those in long-term relationships. After about a year, though, the subjects' NGF levels fell back to a normal level.[1]

Another study, conducted at the University of Pisa, found that individuals who had just fallen in love had elevated levels of cortisol—a hormone often associated with stress. (Love *is* stressful, but cortisol is also released when we are aroused.) After 12 to 24 months, the love study participants were retested and their cortisol levels were back to normal.[2]

It's not necessarily bleak to think that romantic love fades. After all, who could endure the intensity of early love with its insecurity, obsessions, and crazy mood swings? Once the intense phase of intimacy wanes, researchers believe that an affectionate form of love develops, one based on companionship and respect. (Of course, many relationships don't make it past the first rush of love. Nor should they.)

A similar timeline may hold true when it comes to marriage. After the first years of wedded bliss, some discontentment seems to follow. A poll of 5,000 married couples found that men and women begin to take their marriage for granted after two and a half years. Half the couples surveyed for the 2008 study reported that they felt undervalued at the 2.5 year mark. The majority of the men said they stopped picking up after themselves, while the women were no longer making an effort to look nice for their spouse. A 2011 survey of married couples found that irritation peaks at the 3-year mark. More than two-thirds of all of those surveyed said that little quirks, which were seemingly harmless and often endearing during the first flushes of love, became major annoyances at 36 months.

 Book Excerpt: Why sudden love may not lead to lasting love . . .

"If two people who have been strangers, as all of us are, suddenly let the wall between them break down, and feel close, feel one, this moment of oneness is one of the most exhilarating, most exciting experiences in life. It is all the more wonderful and miraculous for persons who have been shut off, isolated, without love. This miracle of sudden intimacy is often facilitated if it is combined with, or initiated by, sexual attraction and consummation. However, this type of love is by its very nature not lasting. The two persons become well acquainted, their intimacy loses more and more its miraculous character, until their antagonism, their disappointments, their mutual boredom kill whatever is left of the initial excitement. Yet, in the beginning they do not know all this: in fact, they take the intensity of the infatuation, this being 'crazy' about each other, for proof of the intensity of their love, while it may only prove the degree of their preceding loneliness."

How long should a person stay in a dating relationship?

Rabbi Yaacov Deyo, the creator of SpeedDating (see box below), maintains that it takes an average of 3 months of dating to determine whether marriage is a viable goal. Deyo explained to me via e-mail, "the idea at work here is that spending time with a person in various situations will give me information, and that once I have enough information I will have the clarity I need in order to determine if this is the right person for me—and vice versa. The mistake many people make is that they believe this principle still applies after three or four months." In fact, long, nonmarital relationships Deyo believes, can be destructive, because they put peoples' hearts at risk. "The partners buy into the trap of mistaking the dating, for the relationship."

Time Trivia: Speed dating

Speed dating may seem crass, but its origins are honorable. The concept was created by Rabbi Yaacov Deyo of Aish Ha-Torah with assistance from his students as a way to help Jewish singles find a life partner. The first event took place at Pete's Café in Beverly Hills in late 1998. Deyo, a Harvard graduate, was ordained as a rabbi at the Aish HaTorah Yeshiva in 1996. SpeedDating, as a single word, is a registered servicemark of Aish HaTorah.

SpeedDating has spawned many copycat operations, but the process typically works like this: At an event individuals meet, round-robin style, in a series of "dates" that last from 3 to 8 minutes. At the end of the event, participants write the names of the singles they liked, and if the other picks them too, a match is made.

Sex and Intimacy

How often do people have sex?

The average frequency is 5 to 7 times a month—slightly more than once a week. A recent *Esquire* magazine survey found that one-third of men have sex 1 to 3 times a week, one-third have sex once a week, and the remaining have sex either much more frequently or much less.

Sexual frequency is dependent on age, marital status, and a few other issues. One recent study of 24- to 45-year-olds, for instance, found that being married increased sexual frequency for women, but not for men. The study also revealed that above-average height, lower education (below high school), and youth were associated with more sexual activity among men (though not among women).[3]

How much time do Americans spend *thinking about* sex?

You've likely heard the urban legend that says guys think about sex every 7 seconds. Wrong. There's no consensus on how often men and women think about sex. However, a recent study tried to shed light on the issue. Researchers gave 283 college students a golf tally counter and asked them to note when they thought about food, sleep, and sex throughout the day for one week. The young men in the study thought about sex *more* than the women, *but* they also thought about food and sleep more than the women. Men had about 18 sexual thoughts per day while women had about 10. That works out to about once an hour for men and once every other hour for women. Men seem to feel more urges of all kinds than women. Are men more primal than women? Or are women more reluctant to state their urges? The authors of the study theorized "it is possible that at least part of this sex difference is due to a greater reluctance on the part of women to report such cognitions. In light of the stereotype that men think much more often about sex than do women, some young women, particularly those with greater discomfort with sexuality, may be reluctant to report that they think about sex very often." Ask your friends and form your own opinion.[4]

World Record: The Longest Kiss

At press time, the record holders for the longest continuous kiss were a gay Thai couple who smooched for 50 hours, 25 minutes, and 1 second starting on Valentine's Day 2012. The lip lockers were not allowed to sit down and were required to take bathroom breaks together. They were permitted to sip drinks through a straw, but their lips could not part. Romantic? Not to me. Lucrative? You bet! The winners won a $3,300 diamond wedding ring and a $6,600 hotel voucher.

How long does a hug last?

About 3 seconds. Researchers analyzed 188 spontaneous embraces at the 2008 Summer Olympics Games between athletes and their coaches, teammates and their rivals, and found the hugs lasted on average for 3 seconds. The researchers also found that the duration of the touch involved in the preparation and release of the embrace lasted an additional 2 seconds. Other studies have found that both animals and humans tend to do many things in 3-second episodes. The author of the hug study, Emese Nagy, a developmental psychologist, said, "Crosscultural studies dating back to 1911 have shown that people tend to operate in 3-second bursts. Goodbye waves, musical phrases, and infants' bouts of babbling and gesturing all last about 3 seconds."[5]

How long does sex last?

Adults spend 16.9 minutes on foreplay and 19.2 minutes, on average, having sexual intercourse, according to Durex's Sexual Wellbeing Global Survey, which was conducted in 37 countries in 2011. See the table below for the details on how countries differ.

A study of Czech women came up with similar numbers. The study, published in 2009 in the *Journal of Sexual Medicine,* found that women spend 15.4 minutes on foreplay, and 16.2 minutes on intercourse.

Esquire magazine found in a 2012 survey that 42 percent of men had sex for 15 minutes or longer, 34 percent for 8 to 14 minutes, and 16 percent for 4 to 7 minutes (just 4% of respondents said less than 4 minutes).[6]

Time spent on sex, by country (courtesy of Durex)

Country	Foreplay	Sex	Total time
Brazil	21	27	48
Greece	21	25	47
Switzerland	20	26	46
New Zealand	19	25	44
Hong Kong	13	29	43
Mexico	21	22	43
Portugal	18	23	41
Columbia	18	22	40
Indonesia	15	24	39
South Africa	18	22	39
Poland	19	21	39
Russia	16	22	37

Hungary	17	20	37
Malaysia	13	24	37
Austria	18	19	37
Netherlands	19	18	36
China	15	20	36
Australia	18	17	35
Croatia	16	19	35
India	19	15	34
Romania	16	19	34
Ireland	17	17	34
Turkey	17	17	34
Italy	17	17	33
Germany	17	16	33
Singapore	15	18	33
Czech Republic	17	16	33
Spain	18	15	33
Nigeria	16	17	33
UK	18	15	33
USA	16	16	33
Japan	17	15	32
Thailand	11	21	32
France	17	14	32
Canada	16	16	32
South Korea	15	16	31
Taiwan	14	17	31
Global Average	**17**	**19**	**36**

How long *should* sex last?

A Canadian study asked heterosexual couples what the ideal amount of time would be for sex and foreplay. Whaddya know? The subjects' answers turned out to be close to the averages cited above. The responses of the women and men were somewhat similar (though men did hope for longer intercourse sessions): women said foreplay should last 18.93 minutes and sex 14.34 minutes; men said foreplay should be 18.10 minutes and sex 18.45 minutes.

Sex therapists see things a bit differently. According to a survey of American and Canadian sex therapists, the optimal amount of time for sexual intercourse is 7 to 13 minutes. The therapists deemed 1 to 2 minutes too short, and 10 to 30 minutes too long, according to the report published in the *Journal of Sexual Medicine* in 2008. It seems that most of the world, then, is spending a few too many minutes on sex.[7]

How long does an orgasm last?

On average a matter of seconds, but there isn't reliable scientific data on this topic. In fact, researchers can't even agree who has longer orgasms, men or women—which is understandable since orgasms are not easy to measure. When do they begin, and when precisely do they end? According to the scant research, the length of an orgasm is somewhere between 3 and 20 seconds and up to one-third of women may experience exceptionally long orgasms that last up to 2 minutes.

Tim Ferriss claims that a woman's orgasm can last for many minutes. In his book *The 4-Hour Body*, Ferriss says that with the right technique it's possible to help a woman achieve a 15-minute orgasm. That's right—15 minutes. The author and entrepreneur knows, because he developed the technique. Buy Ferriss's book and see for yourself.[8]

Sex: the thing that takes up the least amount of time
and causes the most amount of trouble.
—John Barrymore, American actor

How long does it take to reach orgasm?

There's more agreement on this matter. It takes men 2 to 20 minutes of intercourse and women 20 minutes or more to climax. I doubt this comes as much of a surprise to men and women the world over.

What's the best time for sex?

According to a poll of 5,000 Brits, the most popular time for sex is 10:16 P.M. on Saturday night. The runner-up time is 9:00 P.M. on Friday. Third? 9:30 A.M. on Sunday morning. While these results may sound a bit dreary to single folks, bear in mind: couples are likely too tired and cranky during the week to let lust trump their need for slumber (especially if they have kids). Or perhaps they are too tired and cranky to even feel lustful. The study also found that 90 percent of the time, sex takes place in the bedroom.

The poll findings were corroborated by a more scientific study conducted by the Circadian Rhythm Laboratory at the University of South Carolina. The majority of sexual encounters among the study's participants ages 18 to 51 took place at bedtime (11 P.M. to 1 A.M.). But the study also found that 20 percent of individuals reported their usual time for sex was first thing in the morning.[9]

What's the best age for sex?

An online poll asked adults when they experienced the best sex of their lives. Women, on average, said 28 and men said that 33 was the best age. Once again, women seem to mature more quickly than men. Incidentally, 28 to 33 is about the time when reproductive faculties are at their baby-making peak. But an age bias may be at play. When the pollster looked at the results from respondents in their 50s and 60s they found that those folks said their best sex was at age 46.[10]

How long does it take to get over a broken heart?

Broken hearts, like love, can last for days or a lifetime. A study of college students who had been rejected by their partners found that the lovelorn thought about their lost lovers 85 percent of their waking hours. The students' relationships had broken up on average 63 days earlier. The study, conducted by Helen Fisher of Rutgers University, also found—reassuringly— that this intense longing waned over time.[11]

 Time Trivia: More Americans are "living in sin" than ever before.

The percentage of women who were living with a man in a sexual relationship rose from 3 percent in 1982 to 11 percent in 2010.[12]

Love will make you forget time and
time will make you forget love.
—Gordon Livingston, M.D.

Marriage

How long do people wait to get married?

Men delay tying the knot until they are 28 years old, women until 26—the highest ages ever recorded. But the numbers have bounced around over the centuries. In 1890, the first year for which the Census Bureau has data, men got hitched at 26 and women at 22. Those numbers *fell* until the early 1950s—in 1950 men were 23 when they married, women were 20—since then the marriage age has been steadily rising.[13]

Age varies by region: northeasterners generally marry later, while westerners marry earlier. In Rhode Island, for example, women were a median of 29 years old and men, 31, when they first walked down the aisle. In Idaho, by contrast, women were just 23 years old and men, 26.

Time Trivia: Teen Brides

In less-developed countries, girls tend to marry young. One girl in seven marries before age 15, and 38 percent of girls marry before age 18, according to The Girl Effect organization. By contrast in a highly developed country like Switzerland, women wait until they are 32 and men until they are 35 to get married.

How long are engagements?

In the United States, the average engagement—from the day of the announcement to the day the couple walks down the aisle—is 14 months, according to theknot.com, an online wedding site.[14]

Time Trivia: Wedding Prep

In the three months leading up to the big day, brides report they spend an average of 11 hours a week on wedding planning. Given the time a wedding requires, it's not surprising to hear that half of brides surveyed said planning their nuptials was more stressful than they had expected. They tend to share their stress, hence the term Bridezilla.

When do couples get married?

Just over half (53%) of weddings take place in the afternoon, 31 percent occur in the evening, and 16 percent happen in the morning.[15] The most popular wedding month is still June, while the most popular engagement month is December. Why December? It's the month when families are together, and it's also the last month of the year (as in, "we must get engaged by year end, or else!").

Married Life

How much time do married couples spend together?

Less than they used to. In the last two decades of the twentieth century, marriages became more separate, according to a large study by Penn State sociology professor Paul Amato and published in his book *Alone Together: How Marriage in America Is Changing*. The number of couples who spend time eating, doing projects around the house, and visiting friends together declined significantly from 1980 to 2000.

OK, but is that a bad thing? About half of all married folks (or those living with a partner) say their relationship is closer than that of their parents, and 43 percent of couples say their relationship is the same as their parents, according to a 2010 survey by the Pew Research Center. Only 5 percent of couples said their relationship was less close than their parents' marriage. Maybe time together does not necessarily bring couples closer together.

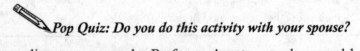 *Pop Quiz: Do you do this activity with your spouse?*

According to a survey by Professor Amato, couples spend less time doing projects together than they used to. See how your life compares to the survey results below.[16]

% of people that reported almost always doing the activity
with their spouse

Activity	1980	2000	Your Answer Here
Work together on projects around home	43%	34%	_____
Eat the main meal together	78	66	_____
Go out for leisure	62	44	_____
Visit friends	53	34	_____

Love seems the swiftest, but it is the slowest of all growths. No man or woman really knows what perfect love is until they have been married a quarter of a century.
—Mark Twain

How much time do couples spend arguing?

A survey of 2,000 married couples found that the newly-in-love bicker for just 1.2 hours a week. Couples who had been together for more than 3 years rowed for an average of 2.7 hours per week.[17] Not so bad. Another survey looked at bickering in terms of incidents, rather than total hours, and found that couples argued about 7 times a day! But if each interaction was a few minutes, the two surveys were not so far off.[18]

Does desire change over time?

Yes, but the change appears to be different for women and men. One large study published in 2002 looked at the love lives of heterosexual German students between the ages 19 and 32 who were in a steady relationship. The researchers found that over a 3-year time span the men had just as much sexual desire for their partner as when they first met, but the women's longing for sex had dropped off significantly. Conversely, women experienced a greater desire for tenderness from their mate, while the men had less.

Of course, this is a single study of students no older than 32 years, but the sample was large and the results intriguing. The findings square with many of the anecdotal reports one hears about married life.[19]

Year 1 and year 3: how desire changes

	During the first year of the relationship	After the third year of the relationship
♂ Desire for frequent sex by men	76%	76%
♀ Desire for frequent sex by women	65%	↓26%
♂ Desire for tenderness by men	70%	↓48%
♀ Desire for tenderness by women	89%	↑93%

Marriage + Longevity

How long do marriages last in the United States?

Marriages that end in divorce typically last 8 years (second marriages that end in divorce last only slightly longer). But when you look at the time from marriage to separation, the time is shorter—about 7 years.

What about happy unions—how long do they endure? If you group together all the couples in the country in first-time marriages, you'll find their unions are a median 21 years long (which means that half have been together more than 21 years, and half for less).[20]

Here's another way to look at how long marriages last: two-thirds of first marriages last 10 years or more.[21]

And another: slightly more than half of all first marriages will reach their twentieth anniversary.[22]

How long do celebrity marriages last?

Kim Kardashian's infamous union lasted just 72 days. And she has plenty of company. The shortlived Kardashian/Humphries marriage might have been predicted by the Sundem/Tierney unified celebrity theory, an equation for predicting the odds that a celebrity marriage will last. Garth Sundem, a self-professed übergeek, and John Tierney, a writer for the *New York Times,* came up with the theory in 2006 and tweaked it in 2012. The new and improved equation states that a few key variables predict whether a fame-filled union will last and for how long: the spouses' combined age (just like in real life, younger couples divorce sooner), the length of the courtship (shorter courtships yield shorter unions), and the sex-symbol factor of the wife, which is defined as "the number of Google hits showing the wife in clothing designed to elicit libidinous intent," explains Tierney. (Kim's "factor" is sky-high.) It's unclear why a woman who seeks constant media attention spells doom for a marriage, but one plausible explanation may be that narcissistic women who need constant fixes of attention can't get those fixes in a stable, long marriage. Needy women may become restless and cheat, or attention-craving creatures who are hell to live with.

How well does the theory work? At press time, the theory was predicting long unions (more than 15 years) for Kate and Prince William, Calista Flockhart and Harrison Ford, Chelsea Clinton and Marc Mezvinsky, and Beyoncé Knowles and Jay-Z. How will the theory hold up? Time will tell.[23]

How long do first marriages last elsewhere?

If you live in Finland, Hungary, the Czech Republic, or another European country, chances are your first marriage will last longer than the U.S. average of 8 years, but these numbers may be influenced by local laws. "In many European countries including Austria, the Czech Republic, France, Germany, Iceland, Ireland, Italy, Norway, Poland, the Slovak Republic, Spain, Switzerland, and the United Kingdom, there are regulations regarding the minimum period that spouses must have lived apart in order for a divorce to be granted," says the Organisation for Economic Co-operation and Development, which compiled the following data.[24] So a couple may have split up after 8 years, but in reality they must wait until, say, the 10-year mark before they can officially divorce.

Average Length of First Marriages in Countries across the Globe

Country	Length in years
Italy	16.8
Slovenia	16.0
Spain	15.2
Bulgaria	14.9
Belgium	14.8
Slovak Republic	14.6
Portugal	14.5

Switzerland	14.4
Poland	14.3
Hungary	14.1
Netherlands	14.1
Czech Republic	14.0
Germany	14.0
Norway	13.6
Luxembourg	13.6
Latvia	13.6
Lithuania	13.5
France	13.3
Estonia	13.2
Finland	13
United Kingdom	13
Greece	12.5
Romania	12.2
Austria	12.1
Iceland	11.9
Sweden	11.5

Denmark	11.5
Cyprus	10.9
Turkey	10.4

What reduces your risk of divorce?

Fewer people are getting married, but those who do stay hitched for longer. How long will *your* union last? You have a better chance of a long marriage if your income is high, you went to college, you married later in life, and your parents were not divorced. The blog *Political Calculations* came up with the following factors that *decrease* your risk of divorce.

Factors that *decrease* your risk of divorce	By what percentage:
Making over $50,000 annually (vs. under $25,000)	–30
Having graduated college (vs. not completed high school)	–25
Having a baby seven months or more after marriage (vs. before marriage)	–24

Marrying over 25 years of age (vs. under 18)	-24
Coming from an intact family of origin (vs. divorced parents)	-14
Having a religious affiliation (vs. none)	-14

 Time Exercise 2: Calculate Your Risk of Divorce

To get a precise read on your own marriage, go to Political Calculations.blogspot.com and plug your vital stats into the calculator "What Are the Chances Your Marriage Will Last?," which uses statistics from the Centers for Disease Control and Prevention. Or try economist and marriage expert Betsey Stevenson's calculator, which you can find at divorce360.com. The calculator can tell you the likelihood that you may divorce in the next five years.

Warning: If both you and your spouse have divorced parents (like me), your prospects will look grim.

World Record: Marriage

How long can a marriage last? For one couple, 91 years.

Daniel Frederick Bakeman and Susan Brewer were married in 1772; he was just 13 years old and she was a mere 14! He lived to 109, and she lived to be 105. What's more, Daniel Bakeman may have been the last surviving veteran of the American Revolutionary War.

After Marriage

How long will it take you to find a new mate?

For people whose first marriages ended in divorce, it only takes a few years to find a new mate. Divorcées get hitched in 3.5 years, divorcés in 3.6.

How long will your second marriage last?

Second marriages that end in divorce don't last much longer than first failed marriages—though Hispanic women have slightly better odds.[25]

Second marriages that ended in divorce	Total	White	Black	Asian	Hispanic
Men	8.5 years	8.7	7.9	NA	7.4
Women	8.0	7.9	8.7	NA	9.0

Is marriage becoming obsolete?

40 percent of Americans say "yes."

Americans used to spend the majority of their adult lives in legal unions. In 1960, 72 percent of adults were married. By 2009, that share had fallen to 51 percent. The United States has one of the highest marriage rates in the Western world, but also one of the highest divorce rates. However, divorce is not fully to blame for the decline of marriage, many people simply choose not to marry. Why? One reason may be that "marriage is increasingly aligned with a growing income gap," wrote the authors of the Pew Research Center's report *The Decline of Marriage and Rise of New Families*. "While declining among all groups, [marriage] remains the norm for adults with a college education and good income but is now markedly less prevalent among those on the lower rungs of the socio-economic ladder . . . those in this less-advantaged group are as likely as others to want to marry, but they place a higher premium on economic security as a condition for marriage. This is a bar that many may not meet."

Ironically, even though many adults believe marriage is becoming obsolete, about half of unmarried adults would like to wed.[26] Perhaps that's because happy marriages are associated with greater health, happiness, and longevity. A happy marriage can be like vitamin B_{12} for your body and soul. Emphasis on "happy."

Family
Moms, Dads & Kids

Americans love family time—or at least they say they do. A full 75 percent of adults claim that their family is the most important part of their lives. The dynamics of family time have changed dramatically over the past few decades, thanks to social and technological revolutions: couples wait longer to get married and have kids, and they have fewer of them; women spend more time at work; parents spend more time with their kids; men do more of the housework; kids spend more time online!

Because men and women now share much of the responsibility for breadwinning, child care, and even parent care, stress is on the rise. Can parents devote time to work, kids, careers, and themselves and still remain sane?

Family Time

How much time do parents spend with their children?

Parents take parenting more seriously than ever. Over the past four decades, moms and dads have gradually increased the time they spend playing with, and caring for, their children, with dads making the greatest percentage gains. Moms—however—still log the most hours of child care, which is understandable since a wife's paid work hours tend to be lower than a husband's.[1]

Child-care hours since 1965

	Fathers	Mothers
2008	7.8 hours per week	13.9 hours per week
2005	6.8	13.9
1995	4.2	9.6
1985	2.6	8.4
1975	2.6	8.6
1965	2.5	10.2

Parents are also spending more quality time with their kids

Parents are giving their kids more praise, structure, and attention than they did just a decade ago, according to Census Bureau data.[2]

- **More time talking and playing.** Today 54 percent of children between ages 6 and 11 talk or play with their parents "just for fun" three or more times a day, up from 45 percent who got such attention in 1998.

- **More praise.** The trend was similarly positive when it came to praise. More than half of children (56%) age 6 to 11 hear praise from a parent 3 or more times a day, up from 45 percent who received such a wealth of kudos in 1998.

- **More TV time rules.** Parents are getting tougher about TV. A large majority (72%) of parents have rules about the number of hours their kids (6 to 11) are allowed to watch TV, the types of programs they can view, and the time of day that the TV can be on. That's an increase from just 65 percent who had such rules a decade ago.

- **More dinner hour togetherness.** Most (75%) of children eat dinner with at least one parent 5 nights a week or more versus 73 percent in 1998. Not a huge jump, but at least the trend is up. (See the dinner-hour challenges in the following box.)

Time Trivia: The imperiled dinner hour

A 2008 study found that dinnertime distractions are a growing problem. More than half (58%) of adults interviewed for the study reported being distracted by technology or entertainment during the dinner hour. Not surprisingly, television topped the list of distracters.[3]

And yet . . . parents feel they don't have enough time with their kids!

Despite their überinvolvement, many mothers and fathers say they have too few hours with their children. "This may reflect the hurried nature of modern family life—when time together is often spent rushing to the next activity or commitment," writes Suzanne Bianchi in her paper "Family Change and Time Allocation in American Families." "The feeling that one has too little time with one's children is much more prevalent among employed than non-employed mothers." Indeed, Bianchi, a professor of sociology at UCLA, found that just 18 percent of stay-at-home mothers felt they had too little time with their children, compared to nearly one-third of part-time employed mothers and over half of full-time employed mothers.[4]

Time Trivia: Which dads log the most hours?

Energetic twentysomething dads spend 4.3 hours each workday with their preteen children. Older dads, those between 29 and 42 years old, spend just 3.1 hours each workday with their kids of the same age.[5]

Dads are more attentive, but fewer of them are present

Overall the stats *look* good: dads are spending more time than ever with their young pups. But drill down into the data and the results are not all that rosy. The percentage of kids without a live-in dad has risen dramatically over the past 50 years. Consider this: in 1960, just 11 percent of kids lived apart from their dads. In 2010, fully 27 percent of kids were not living with their fathers.

Who are these absent dads? A father's race and education are highly correlated with his absenteeism. Black fathers are far more likely to live apart from their children than white fathers (44% vs. 21%); Hispanic fathers are in the middle (35%). Dads who did not complete high school are far more likely to live under a different roof than their kids than dads with a college degree (40% vs. 7%), according to the Pew Research Center.[6]

Wealth, Status, and Family Time

Youngsters whose parents talk to them and expose them to new experiences perform better academically than kids who do not receive such attention. Kids from higher income brackets get more of this crucial parent time than kids from the lower rungs. Children from high-income families receive 400 more hours of literacy activities and spend nearly 1,300 more hours in stimulating places—such as parks, churches, museums, and stores—between birth and age 6 than their low-income counterparts, according to research by Meredith Phillips.[7]

Time Trivia: The cutest age?

The drop-off in adorableness happens to kids early: at age 4.5, right about the time a child is getting ready to head into kindergarten. A recent study[8] asked adults to gauge the attractiveness of children's faces from infancy to age 6. The participants rated infants as cuter than toddlers and toddlers as cuter than older children. The magic number was 4.5; kids above that age were no longer as irresistible as wee ones just a few months behind them.

Family Reading Time

Of all the enriching activities you can do with your children, reading is one of the most important. When you read to toddlers, they develop language and vocabulary skills, along with an appreciation for books and reading.

Once your child is in grade school, it's vital for him or her to become a good reader; by the time kids reach third and fourth grades they need to read well in order to learn. The National

Research Council bluntly states that a child who is a not a moderately skilled reader by fourth grade is unlikely to graduate from high school.[9]

But as children get older, they spend more time online and on cell phones and less time reading books. After age 8, there is a steady decline in the time kids spend reading books for fun, particularly among boys, according to a study by Scholastic Books.[10]

Setting a good example helps. Parents who read frequently tend to have children who read frequently. Letting kids pick their own books also boosts reading time. What's more, teens who said they had interesting books in their homes were twice as likely to be frequent readers as those who said they didn't, according to the Scholastic survey.

Here's a comparison of parent reading time to kid reading time. Kids still read more than their parents, thank goodness.

Reading time /week	Parents	Kids
7 days	12%	17%
5 to 6	16	20
3 to 4	22	22
1 to 2	19	20
Less than once/week	30	21

Working Moms

Employed mothers are fed up with full-time work

Hello, reality: working moms are showing signs of burnout. Of today's employed mothers, 60 percent say part-time work would be their ideal, up from 48 percent who felt that way in 1997. Another 19 percent of working moms say they would prefer not working at all outside the home, according to a survey by the Pew Research Center.[11]

How do *work* schedules affect *home* schedules for women?

No wonder moms want to work part-time: in addition to more hours with their kids, they get more hours of shut-eye. The following numbers are based on women with husbands who work full-time.[12]

Mom's Schedule

Activity	Stay-at-home	Works PT	Works FT
Sleeping	8.77 hours	8.34	8.20
Working	0.07	2.78	5.20
Household activities	3.58	2.62	2.00
Caring for kids	2.60	1.92	1.24
Leisure & sports	4.14	3.43	2.93

Do working moms *like* the way they spend their time?

A landmark study published in 2004 tracked the time a group of working mothers spent on various activities, as well as how they felt while they were doing them. The study, by Nobel Prize winner Daniel Kahneman and four colleagues, produced intriguing data. The activities women enjoyed the most, intimate relations and relaxing, were among the ones they spent the least time on. Ahhh, if only life could be all play and no work.

What's more, women reported that child care was one of the least-favored activities, along with commuting. Hold on: Don't parents wish they could spend *more* time with their kids? Is this just talk? Not necessarily. The authors theorize that because the women were polled on workdays, "they would have been tired, pressed for time, and irritable during the hours they spent with their children." After 8 to 10 hours at the office, moms might rather chat on the phone than assist with homework. The authors continue: "It is unclear what the results would have been if the women had been polled on the weekends."[13]

Weekday activity	Likeability factor*	Percent of time**
intimate relations	4.26	1.6
relaxing	3.71	4.3
socializing	3.59	14.8
eating	3.35	14.7
pray/meditate	3.26	2.8
watching TV	3.17	15.2

preparing food	2.90	7.7
shopping	2.81	3.0
childcare	2.55	7.9
internet/e-mail	2.50	11.8
doing housework	2.39	14.7
working	2.25	40.5
commuting	2.10	10.5

*Subjects were asked to rate activities on a scale of −6 to 6. The figures here, called net affect, are positive scores, minus negative ones.

**Some activities overlap, which is why the figures do not add up to 100%.

Working Dads

Dads would like to work less, too

Most men want to be more involved as parents. More than half (58%) of fathers say they would like to work fewer hours, compared to just 49 percent of men without children.[14]

The more hours men work, the more conflict they feel

And no wonder they want fewer hours. Extra hours on the job are the main contributor to work-family conflict—even more so than the amount of time they spend on child care, chores, and leisure, according to research by the Families and Work Institute. Late nights at the office or work-related travel cause more conflict at home than forgetting to clean up the kitchen.

Hours of work/week	Stress state
50+ hours	60 percent of men say they experience "some or a lot of work/family conflict"
40 to 49 hours	39 percent feel conflicted
less than 40 hours	29 percent feel the pressure of trying to do it all

Work time has become more intense

Men feel they are expected to work harder and faster. In the 1970s, just 52 percent of men agreed with the statement, "At my job, I have to work very fast"; in 2008, 73 percent of men agreed. In 1977, just 65 percent of men said they had to work "very hard" at their jobs; in 2008, that percentage had zoomed up to 88.

The 1970s were no picnic—the country was in a recession and unemployment was high—even so, the workplace was more relaxed. The recent focus on productivity, at a time when both men and women are striving for more work-life balance, has put a lot more stress on working parents.[15]

Chore Time

Who does most of the housework?

Women. Women are doing far less housework than they did back in the '60s, *but* they still do more than men. According to one study (and there are a few), women spend 17 hours on housework each week, down from 32 hours in 1965 when most women did not work outside the home. However, men spend 9.5 hours a week on housework, up from 4.4 hours in the '60s.

Marriage creates more housework for women, but less for men!

Marriage creates 7 hours of housework for women, but saves men 1 hour, according to a University of Michigan Institute for Social Research (ISR) study.

Before women get married they do about 10 hours of housework—such as cooking, laundry, and cleaning—each week. After they get hitched they do 17 hours of housework each week. Men, on the other hand, do 1 hour *less* housework after they get married, or about 9 hours.

The more children, the more work. Women with more than three kids do 28 hours of housework a week, while men with the same number of children do just 10 hours a week.

It's not that men are slouches. "Men tend to work more outside the home, while women take on more of the household labor," said economist Frank Stafford, who directed the University of Michigan ISR federally funded Panel Study of Income Dynamics. "Certainly there are all kinds of individual differences here, but in general, this is what happens after marriage."[16]

What do parents spend their time bickering about? Chores!

According to an online (aka, not too scientific) poll of Brits, the average couple spends 40 minutes a day arguing about household chores. The top triggers include:[17]

1. Leaving clothes lying around the house
2. Putting off home improvements
3. Doing the washing up
4. Not fixing broken household items
5. Not taking the trash out
6. Not making the bed
7. Leaving the toilet seat up
8. Hiding or not owning up to damage
9. Not emptying the dishwasher
10. Doing shoddy DIY

The clash of the multitaskers

Let's forget about chore wars for a moment—they're so twentieth century. Women and men are facing a new battle. Recent research has shown that women are indeed better at multitasking than men. It's not clear why this is so, but women seem able to juggle more tasks at once than men can. But it's a talent that ends up backfiring for working mums. A 2011 study found that working mothers spend ten more hours each week multitasking than working fathers. But for mothers, doing more than one task at a time feels stressful and distressing, whereas for men it does not, it feels A-OK. The differences may have to do with the tasks the parents are juggling. Women tend to multitask by doing two types of housework at once, while men may talk on the phone while taking care of some personal activity. At the end of the multitasking day, women feel more burdened and stressed than men.[18]

Being Mom

Some stats about twenty-first-century mums

- Women give birth for the first time, on average, at age 25.4, up from age 21.4 in 1970.

- Women are more likely to work while they're pregnant. In the postwar years, just 44 percent of pregnant women worked. But between 2001 and 2003, 66 percent of pregnant women worked, and 80 percent stayed on the job until within a month or less of giving birth, according to the U.S. Census.

- Women spend more time in labor (as in childbirth). First-time moms spend more hours in the first stage of labor than they did 50 years ago (2.6 hours vs. 2 hours). Why? Mothers are older than they used to be and heavier—characteristics which can lengthen labor. In addition, moms today are more likely to be given anesthesia, which can slow down the birthing process.[19]

- They also spend more time breast-feeding their babies. More women breast-feed now than ever and devote more time to it—more than 2 hours a day on average. Less educated moms used to be less likely to breast-feed, now women of all backgrounds are about on par. Researchers at Wharton compiled the following table:[20]

Average Time Spent Breastfeeding per Day

Mother's education	<1980	1980s	1990s	2000–05	2006–10
High school diploma or less	1.3 hours	1.7	2.8	3.0	3.1
Bachelor's degree	2.5	2.5	2.8	2.8	3.0
Master's degree	2.6	2.5	2.5	2.6	2.9
Professional degree or Ph.D.	2.3	2.6	2.2	2.6	3.2

 Time Trivia: What time of year are most babies born?

September had the highest number of births in 2010, while Tuesday was the most common day to deliver, according to the National Center for Health Statistics.[21]

Couple Time

Do busy working parents have less intimacy time?

One study says, *No*. Researchers examined the lives of 6,877 married couples to learn what influence work hours might have on sexual frequency. The study found that wives and husbands who spent the most time on household and paid work combined had more frequent sex than couples who spent less time working. Wow!

Are busy people more inclined to have sex? Are sexually active couples more energetic? It's impossible to know for sure, but the authors of the study wrote, "the much-lamented speedup of everyday life and resulting time crunch does not appear to have adverse effects on sexual frequency among our sample of married couples."

The study, "Who Has The Time?," published in the *Journal of Family Issues* also found that women spent 42 hours on average on nine different household tasks each week, while men spent 23 hours on the same tasks. The women spent 19 hours on paid work, the men 34 hours. When work hours were combined, wives spent 61 hours per week on paid and unpaid work, while husbands spent 57 hours, a difference of 4 hours per week. These results are consistent with other studies that find, when you combine both paid and unpaid work, men and women put in almost equal hours.[22]

Extended Family Time

How much time do adults spend taking care of their aging parents?

Government data show that 16 percent of adults were caring for an elderly family member or friend in 2011. On the days that they provided help, the caregivers spent about 3 hours with their elderly charges—those hours do not include travel time. Two thirds of caregivers were providing daily care.[23]

Vacation Time

Vacations are not just for fun, they help families bond and develop a sense of solidarity.[24] Yes, there's often bickering, but most families take the battles in stride. American families take an average of 4.5 trips each year, and most of those are road trips. Moms are most likely to plan family vacations, and they usually start the planning 3 to 6 months in advance.

That said, according to a Trip Advisor survey, 43 percent of travelers with children admit they would often prefer to take vacations without their offspring. The survey also found that 37 percent plan to take a multigenerational family trip in the coming year (perhaps because the extended family can help care for the tots). Family travel time often includes babies, grandparents, and aunts and uncles.

Time Trivia: Making time fly

"Are we there yet?" To avoid this constant question, nearly half of parents turn on a TV or DVD player to entertain their offspring on the road. Just 23 percent encourage their children to read books. If you lean on "screens" during trips, take solace in the fact that most others parents do, too.[25]

Summer Vacation: Good or Bad?

What could possibly be bad about a summer break? Plenty, if you listen to some educators. Most young students lose about two months of mathematical computation skills over the summer months; low-income students also lose more than two months in reading achievement. If you have grade-school-age kids, stock up on flashcards for the summer recess.[26]

Should we take the train to Texas?

How long does it take to get from city to city on foot, by bike, by car, by bus, by train, or by plane? Of course, it's doubtful you're going to walk from Boston to Austin, but the times are intriguing.

Mode of transport	NYC to LA (2,450 miles)	Boston to Austin (1,696 miles)	Chicago to Milwaukee (83 miles)
Feet	38 days, 3 hours	26 days, 4 hours	1 day, 5 hours
Bike	11 days, 17 hours	7 days, 23 hours	11 hours, 10 minutes
Car	41 hours, 40 minutes	31 hours, 13 minutes	1 hour, 41 minutes
Bus	2 days, 21 hours, 55 minutes	1 day, 23 hours, 55 minutes	1 hour, 45 minutes
Train	62 hours	50 hours, 45 minutes	1 hour, 29 minutes
Plane	5 hours	4 hours	1 hour

Book Excerpt: Slow Travel

"It is impossible to be in this high spinal country without giving thought to the first men who crossed it, the French Explorers, the Lewis and Clark men. We fly it in five hours, drive it in a week, dawdle it as I was doing in a month or six weeks. But Lewis and Clark and their party started in St. Louis in 1804 and returned in 1806. And if we get to thinking we are men, we might remember that in the two and a half years of pushing through wild and unknown country to the Pacific Ocean and then back, only one man died and only one deserted. And we get sick if the milk delivery is late and nearly die of heart failure if there is an elevator strike. What must these men have thought as a really new world unrolled—or was the progress so slow that the impact was lost? I can't believe they were unimpressed. Certainly their report to the government is an excited and an exciting document. They were not confused. They knew what they had found."[27]

—John Steinbeck, *Travels with Charley: In Search of America*

Anticipation is the best part of a vacation

According to a Dutch study of 1,530 adults, the time spent anticipating a trip brings more happiness than the trip itself. Researchers compared a group of adults who were about to take a vacation with another group who were staying put. The researchers found that the would-be vacationers were significantly happier than nonvacationers before their vacation, but not after. Anticipating the vacation actually boosted the subjects' happiness for weeks.

The trip itself, though, did not significantly boost the vacationers' mood. "Generally there is no difference between vacationers and nonvacations post-trip happiness," the authors of the study wrote.[28]

The takeaway: plan your vacations well in advance and spend a lot of time thinking and talking about them. What's more, to maximize the anticipation effect, try to take a few small trips a year rather than one long holiday.

Home
Real Estate & Household Items

Home is where the heart is. It's also the place we spend most of our time: sleeping, eating, and hanging out. And despite the mortgage crisis, most Americans still want to own a home. Homeownership is off, but only slightly. Today 66 percent of American homes are owner occupied, down from just 68 percent a decade ago before the crisis hit. What's more, 95 percent of Americans feel it's important for their children to own a home someday.[1]

Whether you own or rent, your home and the things inside it are time sensitive. Most houses will endure for decades, even centuries, but appliances, furniture, and the beauty products in your bathroom have more limited life spans. Appliances need to be replaced, lipsticks need to be tossed, roofs need to be reshingled. Knowing the timeline of your beloved home and the stuff inside it can help you avoid unpleasant surprises and plan for the inevitable ravages of time.

Buying, Selling, and Moving

How long do buyers spend house hunting?

About 12 weeks. In 2011, buyers spent a median of 12 weeks on their house hunt, up from 7 weeks in 2001. Buyers in the Northeast spent the most time searching, compared with buyers in other parts of the country. Today's buyers have more choices and fewer buyers to compete with, so they can afford to take more time in their home search. In addition, thanks to the mortgage crisis, getting approved for a mortgage and getting a home appraised take longer than in years past.

Year	Median search time
2011	12 weeks
2010	12
2009	12
2008	10
2007	8
2006	8
2005	8

2004	8
2003	8
2001	7

Time Trivia: Buyers and Sellers

The median age of first-time home *buyers* is 30; the median age for repeat buyers is 49. The median age of home *sellers* is 49.[2]

What time of year do people move?

The summer. Nearly one-third of movers pick up stakes in June, July, or August. Spring and fall are tied for second place. Winter is the least favored (though not by as much as you may think): 21 percent of movers relocate in December, January, or February.[3]

How long do people stay in their homes?

Americans are staying put longer. In 2011, homeowners spent 9 years in their homes before selling, up from 6 years in the more peripatetic period between 2001 and 2008. Families with young children move more frequently, presumably because they are in search of bigger digs; older folks with no kids at home move less often.[4]

Homeownership Tenure

	Median years in current abode
Total population	9
Income <$45,000	11
$45,000 to $74,999	10
$75,000 to $99,999	10
$100,000 >	9
Northeast	9
Midwest	10
South	8
West	10
Children under 18	7
No Children under 18	11

Life Span of Household Items

.... kitchen appliances[5]

Like most gadgets, appliance longevity depends on how often you use them and how well you maintain them. Some brands simply hold up better than others. Here are general life expectancy rates for kitchen gear.

Appliance	Lifespan
Dishwasher	9 to 13 years
Disposal	11 to 12
Exhaust fan	10
Faucet	15
Microwave	7 to 11
Refrigerator	11 to 14
Range, gas	13 to 15
Range, electric	13 to 16
Sink, enamel steel	5 to 10
Trash compactor	6

Protect your health: regularly toss out key kitchen items

The kitchen is a prime breeding ground for microorganisms. Get in the habit of replacing germ-prone items regularly.

Wooden spoon: Every 5 years, earlier if the woods splits. Cracks can become homes for organisms.

Sponge: Every two weeks. In between, disinfect your sponge by microwaving it in a bowl of water for two minutes.

Cutting board: Toss when the board develops grooves or splits in which nasty germs can take up residence and multiply. And get in the habit of using one board for meat and another for vegetables.

. . . home appliances[6]

Specific life expectancies will depend on the manufacturer, the design, maintenance, and frequency of use. Here are some general guidelines.

Appliance	Life Span
Air conditioner, room	9 to 10 years
Air conditioner, central	7 to 15
Boilers, gas	20 to 40
Boilers, electric	13
Dehumidifiers	8 to 10
Dryer (gas + electric)	12 to 13
Furnace, gas	15 to 18
Furnace, oil	17 to 20
Furnace, electric	15
Humidifier	8
Washing machine	10 to 12

Water heater, gas	10 to 11
Water heater, electric	11 to 13

How long do lightbulbs last?

Short-lived incandescent bulbs are being phased out of existence. And for good reason—they don't last very long and waste lots of energy.

The bulb	The life span
Incandescent	1,000 to 2,000 hours
Compact fluorescent (CFL)	6,000 to 12,000
LED (light-emitting diode)	20,000 to 50,000 (wow!)

. . . the roof[7]

Material	Life Span
Asphalt	15 to 20 years
Clay/Concrete	Lifetime
Coal and tar	30
Copper	Lifetime
Slate	50 to 100
Wood	30

. . . mattresses and pillows[8]

Your mattress will likely last for decades, ditto your pillow—but they may not be comfortable to rest on after years of tossing and turning. Finicky sleepers may want to replace the items they snooze on earlier, rather than later. Here are some considerations:

Mattresses   7 to 10 years

Over time the coils flatten and provide less support for your back. Cheaper mattresses may break down more quickly than high-end models.

Pillows, down and foam   1 to 10 years

Why the variation? Cheap pillows may lose their shape within 12 months, while higher-end pillows, typically those filled with down, may stay supportive for a decade.

Time Exercise 3: A Clean House in Less Than 30 Minutes

According to *Real Simple* magazine, you can make your home clean and presentable in under 30 minutes (a thorough, cut-the-grease cleaning will, alas, take a bit longer). Try the magazine's speed clean routine for yourself.

Kitchen

- Clear out and wipe down the sink (5 minutes). No piles of dirty dishes! Move them into the dishwasher, and wipe the sink with a sponge.

- Wipe down countertops and stove (1 minute). Clean splatters and spills with a damp cloth or sponge and an all-purpose cleanser.

- Wipe problem spots on the floor (2 minutes). You can save a full-floor mop for the weekend, but use the same cloth (once you've finished with the countertops) to quickly clean any spills or sticky spots, which will attract dirt and get more noticeable if left alone.

- Fold or hang dish towels (30 seconds). Even if they're clean, a jumble of dish towels on the counter can look messy. Take a few moments to fold or hang them.

Bathroom

- Wipe out the sink (30 seconds). If you use a premoistened cloth to wipe your face, swipe the sink bowl and faucet handles with it, too. Or use a washcloth, paper towel, or a product like Windex Glass and Surface Wipes, which don't leave streaks on chrome or mirror.

- Clean splatters off the mirror (15 seconds). Got foamy toothpaste spray on the mirror? Do a quick swipe with the same cloth you used on the sink.

- Wipe the toilet seat and rim (15 seconds). Same cloth! Just do the toilet last.

- Swoosh the toilet bowl with a brush (15 seconds). If you see a ring, give it a quick scrub.

- Squeegee the shower door (30 seconds). Wipe down glass doors to remove water droplets that can cause spotting. No squeegee? Use a dry towel.

- Spray the shower and curtain liner with a shower mist (15 seconds). A quick spray with a daily cleanser will reduce buildup of mildew and soap scum.

Bedroom

- Make your bed (2 minutes). Even a fluffy down comforter pulled up over messy sheets will look polished.

- Fold or hang clothes and put away jewelry (4 minutes). Even better: resist the urge to toss them somewhere in the first place! Put them away as you take them off.

- Straighten out the night-table surface (30 seconds). Take last night's water glass to the kitchen, stow your reading glasses in a drawer, and straighten books or magazines.

Living Room

- Tidy the sofa (2 minutes). It's likely the focus of the room, so neaten it. Fluff the pillows and fold the throws.

- Pick up crumbs with a handheld vacuum (1 minute). Concentrate on surfaces in plain sight: sofa cushions, coffee table, and rugs in the middle of the room. Look for dust bunnies, too.

- Wipe tables and spot-clean cabinets where you see fingerprints (1 minute). Use a microfiber cloth or a Swiffer cloth to pick up dust. If the surfaces are streaked or sticky, use a moist cloth.

- Straighten coffee table books and magazines (2 minutes). Toss old newspapers and corral the remote controls into one place (a drawer, if possible).

• Clear major clutter (5 minutes). Stash video games, toys, and anything else you might trip over.

Last, I might add, spritz the air with lemon, tangerine, or even a puff of your favorite perfume—to give your home a fresh, clean aroma.

From RealSimple.com, reprinted with permission of *Real Simple*.

Events are perceivable, but time is not.
—James Jerome Gibson, American psychologist

How much time does a washing machine save?

None! In fact, the washing machine can *make* work. Folks with washing machines at home are apt to spend more time doing laundry than those without, according to researchers. A nationwide laundry habits survey conducted by the Coin Laundry Association, for instance, found that people with washers and dryers at home spend 50 percent more time on laundry each week than those who don't. (Of course, it's in the CLA's interest to state this finding, isn't it?)

Why do washing machines make work?

1. You tend to wash clothes more frequently, because you can!

2. You do more loads at home because your washer is smaller than the jumbotrons at the coin op. When you

go to the Laundromat, you can do a few loads at once cutting down on your overall wash time.

3. You buy more clothes! And you buy more delicate items that require special washing cycles, or that must be tumbled gently and slowly or line dried.

If you feel bogged down by laundry time, you could try the new speed machine that purports to wash 15 pounds of laundry in 12 minutes—but it only works on clothes that are lightly soiled.

. . . bathroom appliances[9]

Appliance	Life Span
Bath cabinets	Lifetime
Faucet	20+ years
Showerhead	Lifetime
Shower enclosure	50
Shower door	20
Toilet	Lifetime
Whirlpool tub	20 to 50

Protect your health: toss key bath items on time

The bathroom, like the kitchen, can become a hothouse for germs. What's more, overused items won't do their job well. Don't be cheap; replace these inexpensive items every 3 to 12 months.

Toothbrush   Every 3 months; after that time, the bristles begin to fray making them less effective at removing plaque.

Toilet brush   Every 6 to 12 months, or when the bristles become worn and unable to effectively scrub away germs.

Contact lens case   Every 3 months or so, unless you wash the case in the dishwasher periodically. Over time, parasites and fungi can take up residence in your case.

Razor blade   After 5 to 7 shaves. Beyond this time the blade can become nicked and either cut your skin or harbor bacteria—Yuck. Gals: If you shave your legs every day and have thick hair, ditch your blade after 3 to 5 shaves.

When should you replace your beauty products?

Frequently. Over time, beauty products lose their potency and healthfulness. When you buy a new product, write the date on the bottom with a sharpie pen—this way you'll know exactly how long you've had it. The times given are for opened products.

Product	Toss after
Body lotion Lotion can become contaminated with bacteria; err on the side of caution and replace annually.	1 year
Sunscreen The protective ingredients decompose over time. Don't risk your skin, buy a new bottle each year.	1 year
Mascara and eyeliners Don't mess with the health of your eyes. Bacteria on brushes or pencil tips can irritate or infect your eyes. Replace your mascara and liners every 2 to 3 months.	3 months
Lipstick Most sticks and glosses will stay infection-free for years. But if you tend to kiss a lot of cheeks (including your pet's!) which harbor microorganisms, be sure to get new lipsticks annually.	1 to 3 years
Foundation Bacteria love water-based liquids. Even if your potion looks safe, be vigilant about updating it.	6 to 12 months

Face powder Because they are dry, powders have more staying power than most beauty products. To keep them that way, store in a non-humid environment—meaning outside the bathroom.	2 years
Perfume The scent will start to change and fade within 24 months.	2 years
Nail polish Old polish won't harm you, but it will get gloppy over time.	2 years
Shampoo and conditioner The active ingredients will begin to break down and become less effective after a couple of years.	2 to 3 years
Mouthwash Ditto	3 years
Deodorant Ditto	1 to 2 years

 Time Trivia: How long would it take for termites to destroy your wood home?

A termite-infested home, left untreated, could be destroyed within 10 to 15 years, but significant damage might occur within 3 years. A colony consisting of about 50,000 workers could munch a 2" x 4" x 12" piece of wood in about 118 days. And FYI: A queen termite can live for up to 100 years.[10]

DIY Time

Do it yourself or hire a pro? How often have you considered that question? Let's face it: unless you're a carpentry wiz, it's usually faster to hire a professional (but almost never cheaper). And some jobs are just too gross to tackle on your own, like cleaning the chimney. (To learn about how long other projects take, plus their costs, check out the site diyornot.com.)

Project	DIY time	Pro time
Build a bookcase (6' x 6')	14 hours	7 hours
Replace a medicine cabinet	5	4
Replace drawer glides	4	2
Replace a bathroom faucet	2	1
Install a pet door	2	1
Build a gazebo	47	30
Build a doghouse from a kit	9	7
Install a ceiling fan	4	1
Hang a flat-panel TV	3	2
Paint a room (15' x 20')	10	6
Clean the chimney	2	1
Clean 500 square feet of carpet	3	1

How long does it take to build a house from a kit?

It takes about 350 hours to put together a Sears kit house. A *what* house?

Enterprising manufacturers came up with the idea to sell prefabricated homes by mail to the country's booming middle class in the early 1900s. Sears Roebuck sold an estimated 100,000 kit houses to buyers across the United States between 1908 and 1940. The kits were inexpensive and relatively quick to assemble. Sears promised that a precut house with fitted pieces would take only 352 carpenter hours to assemble as opposed to 583 hours for a conventional house. That's about 44 days of work (or 6 weeks—if you worked on the home for 8 hours a day, 7 days a week), as opposed to 73 days.

You can still buy kit homes. There are kits for log homes, timber-frame homes, domes, or panelized houses, according to *Mother Earth News.* These kits can be basic starter homes or mini-mansions. How long do the new homes take to build? A simple log cabin can take as little as 9 weeks to put together. A more conventional home will typically take from 12 to 16 weeks.[11]

Books

How long should you keep books?

As long as they continue to serve you. Some people love to hang on to every book they've ever read; others like to keep just the titles that are meaningful. Books are reminders of magical reading moments, but they take up a lot of room and can attract dust mites. If you struggle with this question (as I certainly do), consider the following opposing points of view from two of the book-loving authors who gave their opinions to the *New York Times* in 2009.[12]

Pare your collection to the essentials

Billy Collins served as poet laureate of the United States from 2001 to 2003:

"For anyone attached to the book as an object of beauty or to one's own library as a physical testimony to the depth and breadth of one's literary experience, such shedding requires a certain ruthlessness.

"But once I decided to simplify the process by keeping only books I was sure to open again, I was amazed at how many books suddenly fell on the dispensable side of that dichotomy.

"Frankly, I am well into the second phase of life when one begins to enjoy getting rid of all the stuff one enjoyed accumulating in phase one. And who needs such elaborate announcements of one's literary credentials?"

Keep them all

Joshua Ferris is the author of the novels *Then We Came to the End* and *The Unnamed:*

"Get rid of books? Are you kidding? The only reason anyone should get rid of a book is if they're going for that Japanese minimalist design look in which the room is all white and not even the drawers are visible. For those of us with more modest decor goals, living everyday lives with clutter and old clothes, cats and children, sour towels hanging from the rack, knickknacks, pilled throws, boring old mementos, what could be more essential than books?"

RIP: For how long did these popular gadgets and cars survive?

Sony Walkman: 31 years (1979 to 2010)

The revolutionary device was reputedly created to help Sony's cochairman, Akio Morita listen to operas during his trans-Pacific plane trips. The idea caught on, big time: Sony sold 200,020,000 of its cassette-based Walkmans. The majority of sales were before the year 2000—after that iPods and MP3 players rendered the boxy Walkman a quaint relic.

Commodore 64: 12 years (1982 to 1994)

The 8-bit computer, which retailed for just $595, was a huge success. About 17 million units were sold during the computer's brief lifetime. Why so popular? The C64 could be plugged right into your home TV set, and was an excellent vehicle for playing computer games.

Kodachrome color film: 74 years (1935 to 2009)

Immortalized by Paul Simon in his song of the same name, Kodachrome was the first popular color film and was beloved by photographers for its realistic tones and rich colors, as well as its durability.

Betamax: 27 years (1975 to 2002)

Sony came out with the first home videotaping equipment in 1975. JVC came out with its own equipment and format a year later. While JVC licensed its VHS format, Sony would not license Betamax. Sony's decision was a death sentence; even though many experts deemed the quality of Betamax tape to be higher than VHS, the brand never gained momentum.

8-track tape: 40+ years (1964 to ?)
William Powell Lear, the creator of Lear Jets, is considered to be the inventor of the first commercial 8-track tapes. In 1966 all Ford cars offered a factory-installed, in-dash 8-track player, making 8-track the most popular car stereo format in the 1970s. The major record labels stopped supporting the format in the mid-1980s, but some entrepreneurial music lovers continue to release music on 8-track.

Saturn: 19 years (1991 to 2010)
Launched as a "different kind of car company" by General Motors, the Saturn brand was initially a hit with consumers. But the brand was not well managed and never became profitable. "Saturn was killed by its creators, GM and the UAW," wrote Paul Ingrassia, author of *Crash Course: the American Automobile Industry's Road from Glory to Disaster,* in an opinion piece in the *Wall Street Journal* on October 2, 2009. "The company starved Saturn for new products, and the union waged war against Saturn's labor reforms to keep them from spreading to other GM factories."

Food

How long does food last?

Now on to the food in your fridge and pantry—how long can things last with becoming gross or poisonous?

Some foods have a longer life span than *you* (salt, whiskey, even honey can last for decades), but most are time sensitive (meat, fish, vegetables). Should you wait for mold to appear or try to stay ahead of the game? Since not all foods show visible signs of spoilage, best to memorize some of the key times below, which come from the incredibly helpful website stilltasty.com.

Temperature is important. Invest in thermometers for your freezer and refrigerator to ensure your food stays at these optimum temperatures: fridge, 35 to 40°F; freezer, 0°F or lower.[13]

 Time Tip: Don't leave food out for longer than two hours; use or freeze all leftovers within four days.

The Food	pantry	fridge	freezer*
NUTS			
Almonds and Peanuts (opened jar or bag)	1 month	4 to 6 months	9 to 12 months
Walnuts (out of the shell)		6 months	1 year
Almonds and Peanuts (unopened jar or bag)	1.5 to 2 years		
Peanut butter (open jar)	3 months	3 to 4 months (after pantry storage)	
Peanut butter (open jar, natural)		6 months	

DAIRY			
Whole milk		1 week after sell-by date	3 months
Butter	Butter can be left at room temp for just 1 to 2 days	1 month after sell-by date	6 to 9 months
Eggs, fresh		3 to 5 weeks	1 year (do not freeze eggs in their shell!)
Cheddar cheese, block, unopened		6 months	6 to 8 months
Cheddar cheese, block, opened		3 to 4 weeks	6 months
Cream cheese, opened		1 to 2 weeks	2 months
Sour cream (unopened)		10 to 14 days after sell-by date	
Sour cream, opened		2 weeks	
BEANS			
Black beans, canned, unopened	2 to 5 years (after that, texture and flavor may change)		

Black beans, canned, opened		3 to 4 days	1 to 2 months
Black beans, dried, in a bag	1 year		
MEAT			
Steak, fresh		3 to 5 days	6 to 12 months
Hamburger, fresh		1 to 2 days	3 to 4 months
Hot dogs, unopened		1 week after sell-by date	1 to 2 months
Hot dogs, opened		7 days	1 to 2 months
FISH			
Can of tuna or sardines (unopened)	3 to 5 years		
Can of tuna or sardines (opened)		3 to 4 days	3 months
Smoked salmon, sliced at counter		1 week	2 months
Fresh salmon		1 to 2 days	2 to 3 months
POULTRY			
Fresh chicken, raw		1 to 2 days	9 months
Chicken salad		3 to 5 days	
Turkey cold cuts, sliced at counter		3 to 5 days	1 to 2 months

CONDIMENTS			
Mayonnaise, opened		3 to 4 months after date on package	
Mustard, opened	1 to 2 months	1 year	
Ketchup	1 month	6 months	
Marinara sauce, jarred, opened		7 to 10 days	4 to 6 months
Marinara sauce, jarred, unopened	12 to 18 months		
Horseradish, opened		3 to 4 months	
Olive oil, opened	18 to 24 months	18 to 24 months	
Maple syrup, 100% pure	Unopened, 1 year	Opened, 1 year	Unopened, indefinitely
GRAINS			
Brown rice (uncooked)	1 month	18 months	18 months
Brown or white rice, cooked		5 to 7 days	6 months
White rice (uncooked)	Indefinitely	Indefinitely	Indefinitely
White flour (unopened or opened)	1 year	2 years	2 years

Whole wheat flour (unopened or opened)		6 to 8 months	2 years
Pasta, uncooked	3 years		
Pasta, cooked		3 to 5 days	1 to 2 months

*Most foods that are kept frozen at 0 degrees will keep indefinitely, but their taste and texture will change.

 Time Trivia: The longest-lived sandwich

How long does a sandwich last? In general, a couple of days. But the Booker chain in the United Kingdom has invented a sandwich that will stay fresh and tasty for 14 days. It's all in the packaging. Booker is using a process called "gas-flushing" that replaces oxygen with carbon dioxide and nitrogen inside the barrier packaging. In addition, sandwich fillings are mixed with a slightly acidic mayonnaise, which acts as a preservative. Some of varieties are distinctly British: cheese and onion, and chicken tikka.

How can you tell when an egg is past its prime?

Place eggs (still in shells!) in a bowl of cold water.

- If an egg lies peacefully on the bottom of the bowl, it's fresh

- If an egg stands on its pointed end, it's still safe but best used for baking.

- If it floats, toss it out.

How does boxed milk last *so long*?

Because of the way it's processed. Shelf-stable milk (as it's known in the dairy industry) has undergone ultra-high-temperature processing (UHT), which heats milk to 280°F for 2 to 4 seconds to kill off bacteria. UHT milk doesn't need to be refrigerated and stays fresh for 6 months.

Regular milk undergoes pasteurization, which heats milk to either 145°F for 30 minutes or 160°F for at least 15 seconds. Pasteurized milk typically stays fresh for 2 weeks, but must be kept cold.

FYI: Skim milk does not have a noticeably different shelf life than regular milk.

 Time Trivia: Which soda takes the longest to pour?

Diet Coke, according to Heather Poole, an airline attendant and author of *Cruising Attitude*. "Of all the drinks we serve, Diet Coke takes the most time to pour—the fizz takes forever to settle at 35,000 feet. In the time it takes me to pour a single cup of Diet Coke, I can serve three passengers a different beverage," Poole told *Mental Floss* magazine.

Whatever begins, also ends.
—Seneca, Roman philosopher, mid-first century A.D.

Body
Mind, Drugs & Digestion

The human body is always at work: creating, renewing, and repairing. It manages to do its work with nearly perfect timing. Our hearts beat, our eyes blink, our breath goes in and out. "We are survival machines," wrote scientist Richard Dawkins in his book *The Selfish Gene*. "Robot vehicles blindly programmed to preserve the selfish molecules known as genes. This is a truth which still fills me with astonishment."

How long does it take to grow an inch of hair, digest last night's dinner, memorize a deck of cards? Behold here, the wondrous times of the human body.

Body Times

How long does it take for your body to . . .

Make a pint of blood?

If you donate (or lose) a pint of blood, your body replaces the fluid within 24 hours, replenishes the lost red blood cells in 2 to 4 weeks, and restores iron to its original level in 8 weeks, according to the Red Cross.

Mend a broken bone?

The bulk of the repair work happens in 6 to 8 weeks, depending on the bone and the severity of the break. But full recovery can take months, even years. Bones heal in three stages: the inflammatory stage lasts about 1 week; the repair stage lasts 4 to 6 weeks; the final remodeling stage where the bone is restored to its original shape and strength, can take 3 to 5 years! *But* the bone is typically restored to adequate strength in 3 to 6 months.

Create a new skeleton?

The body is constantly upgrading its skeleton—so you always have a mix of old and new bone. It takes about 10 years for the body to completely renew the skeleton.

Grow an inch of hair?

The hair on your head grows an inch every 2 months. Each hair follicle goes through three stages of growth: anagen, catagen, and telogen. During the anagen, or growth, phase, the hair grows for about 6 years (eyelashes, in contrast, only last

5 months). The follicle then moves into the catagen, or transitional, phase for 2 weeks. During the last stage, called telogen or resting, the follicle chills for about 1 to 4 months until a new hair pushes the old one out. The majority of the hairs on your head are in the anagen phase.

There are exceptions to every rule: Xie Qiuping of China must have a very long anagen phase; her ponytail measured 18 feet, 5 inches long, which gained her a spot in the *Guinness Book of World Records*.

Grow an inch of nail?

You gain an inch of fingernail in about 10 months. Your nails grow faster in the summer than in the winter. Toenails grow slower than fingernails, possibly because there is less blood circulating in toes than in fingers.

Create new taste buds?

Each taste bud in your mouth is renewed every 7 to 14 days. Your tongue has about 9,000 taste buds—each bud comprises 50 taste cells.

Create a new layer of skin?

Every 2 to 4 weeks you have a completely new layer of outer skin, or epidermis. About 50,000 dead skin cells fall off your body every minute. You lose 2 ounces of skin (yuck) per month.

Make a pint of mucus?

A matters of hours. Your nose makes 2 to 3 pints of mucus a day. Sound like a lot? Most of the mucus ends up trickling down your throat rather than out of your nostrils.

Repair a damaged cornea?

Twenty-four hours. The cornea is the only part of the eye that has the ability to constantly renew itself. Isn't the body amazing? If our corneas were not so able, we might go blind at an early age.

Heal a wound?

It takes 6 to 18 months for a stitched wound to completely heal. FYI: rubbing creams or lotions on a scar won't hasten the healing process.

 Time Trivia: Are humans still evolving?

Some scientists say yes. As our lives have become more sedentary, for instance, our bodies have become less muscular and our bones have become thinner. Studies of ancient skeletons reveal that around 2 million to 5,000 years ago, our bone strength fell by 15 percent; over the next 4,000 years, bone strength fell another 15 percent.[1]

A person will sometimes devote all his life to the development of one part of his body—the wishbone.
—Robert Frost, American poet

When is the best time to . . .

see the dentist, write a novel, or make love?

Your body is in a constant state of flux. Throughout the day, your temperature, hormone levels, and blood pressure rise and fall; your organs go from more efficient to less so; and your muscles become more or less pliable. To get the most from your body, consider these timing tips.

Have surgery:	8 A.M. Your clot-forming platelets are at their stickiest.
See the dentist:	Afternoon. Your dental pain threshold is at its peak
Have a cocktail (or two):	Between 5 and 6 P.M. (happy hour!). The liver is most efficient at detoxifying alcohol.
Get a gal pregnant:	Afternoon. Semen is friskiest.
Solve problems:	10 A.M. Your mind is most alert.
Lift weights:	5 P.M. Your muscles are at optimal strength and most able to do heavy lifting.
Turn on the air purifier:	At night. Your airways are most constricted at 5 A.M., which is why asthma attacks occur most frequently between 4 A.M. and 6 A.M.

Get tested for allergies: Evening. Results of skin testing for allergies are more pronounced in the evening than in the morning.

Avoid stress: 7 A.M. Blood pressure peaks and sudden cardiac death is most likely to occur in the early morning. This is *not* a good time to engage in intense or stressful conversations.

How long does it take for a dental cavity to form?

Cavities, also known as caries, can form in a few months or a few years depending on your diet, how often you brush and floss your teeth, and whether you live in an area with fluoridated water. If you do get a cavity, you need to have it filled.

 Time Trivia: Longest hiccup attack

The world's longest hiccup attack lasted 69.5 years and was endured by American Charles Osborn; the longest sneezing attack dragged on for 978 days and was borne by Donna Griffiths of the United Kingdom.

How long do common ailments last?

Most minor ailments, like colds, clear up in a matter of days. But basic infections can morph into complicated maladies that take weeks to heal. A common cold can escalate into bacterial pneumonia, which can impair the lungs for a month or more. What's more, one cold often begets another. If you feel sick all winter, it's likely that you no sooner kick one virus than another comes to roost. Reminder: antibiotics only help with bacterial infections, *not* viral ones—like colds and flus. Only take antibiotics when you truly need them.

Infection	Duration
Cold	3 to 5 days, but in 25 percent of patients, colds last up to 2 weeks
Flu	2 to 5 days, on average; in some cases, the flu can linger for a week or longer
Bronchitis	3 weeks, though the cough can linger for months
Sore throat	1 week or less
Sinus infection (acute sinusitis)	3 to 4 weeks, if untreated
Strep throat	2 to 3 days (after you start taking antibiotics, which are typically necessary)
Stomach bug (gastroenteritis)	1 to 2 days, though some bugs may linger for up to 10 days

How long can germs live outside the body (like on your countertop)?

The nasty bugs that cause colds, flus, and food poisoning can survive longer than you might expect. Viruses, like those that cause influenza and SARS, can only last a few days on a dry surface, while some bacterium, such as E. coli, can survive for months, according to a review published in the journal *BMC Infectious Diseases* in 2006. Viruses need a host, such as your cells, to reproduce, so their life spans outside the body are typically shorter than that of bacteria, which can reproduce on their own.[2]

How long does it take to digest your dinner?

Foods high in fat and rich in protein take longer to digest than low-in-fat foods and carbohydrates. Whole milk, for instance, takes longer to make it through your system than, say, a banana or a bowl of rice.

After you eat a meal, it takes about six to eight hours for the food to pass through your stomach and small intestine. What's left of your meal then enters your large intestine (colon) for further digestion and absorption of water. Complete elimination of a burger and fries dinner may take several days.

In the 1930s, two English doctors did some enterprising research on time and digestion. The docs looked at how long it took for specific foods to leave the stomach, a process called "gastric emptying." Modern-day scientists, alas, have not replicated their research. The results of the British doctors' experiments, shown below, are valid, but not necessarily precise. Take note of the quirky food choices, such as boiled beef and stewed gooseberries, which were favored victuals in 1930s England.[3]

Food	Time it took for the food to leave the stomach
2 hard-boiled eggs	4.5 hours
Banana	4.25
Beef, boiled	4
Honey	4
Potato, boiled	4
Gooseberries, stewed	3.75
White bread	3.75
Whole milk, ½ pint cold	3.5
Whole milk, ½ pint boiled	2.5
Skim milk	2.5
Vegetable salad	2.5
Honey dissolved in ½ pint of water	1.5
Pint of water	.75

How long does it take to digest a cocktail?

It takes about 1 hour for the liver to break down the alcohol in a typical drink. To avoid feeling tipsy, sip your beverage slowly. When you knock drinks back at the rate of more than one an hour, your liver can't handle the load and the excess alcohol goes into your bloodstream and your brain—resulting in drunkenness, impaired judgment, and possibly embarrassing behavior.

How quickly do drugs take effect?

Some medications start to work in seconds while others can take days to be effective. The following chart answers questions about some of the most popular twenty-first-century drugs: prescribed and illicit.

Half-life is the time it takes for your body to metabolize and excrete half the ingested substance. Why is this number important? "Half-life is one factor that helps to predict how frequently medications should be taken to be most effective," explains Dr. Amy Sapola, a pharmacist with the Mayo Clinic. For example, some meds spend longer in the body and so do not need to be taken frequently, like Valium. Others, like ibuprofen, need to be ingested more often for continued effect. How long does it take for a medication to *completely* clear your body? A single dose of most drugs will be gone in 4 to 5 times the half-life, says Dr. Sappola (who also graciously fact-checked my numbers).

A note about the data: The chart data assume that one dose of the drug was given to a person in good health.

Prescribed drugs

Brand name (generic)	When it starts working	How long it keeps working (also known as duration of effect)	Half-life
Ambien (Zolpidem)	3 to 15 minutes	2 to 4 hours	2.5 hours
Cipro (Ciprofloxacin)	<1 hour	12 hours	5 hours
(Ibuprofen)	25 to 30 minutes	4 to 6 hours	2 to 4 hours
Tylenol (Acetaminophen)	30 minutes	6 to 8 hours	2 to 3 hours
Ritalin (Methylphenidate)	30 minutes	3 to 6 hours	2 to 4 hours
Valium (Diazepam)	15 to 30 minutes	12 to 24 hours	20 to 70 hours!
Xanax (Alprazolam)	15 to 30 minutes	Up to 6 hours	11 hours

Abused drugs

Cocaine (Cocaine hydrochloride)	Within seconds	1 to 2 hours	30 to 90 minutes
Ecstasy (MDMA)	20 to 60 minutes	2 to 4 hours	8 hours
Heroin (Diacetylmorphine, injected)	1 to 2 minutes	3 to 5 hours	15 to 30 minutes
LSD (Lysergic acid diethylamide)	20 to 60 minutes	Up to 12 hours	3 to 5 hours
Marijuana (Cannabis, smoked)	6 to 12 minutes	2 to 6 hours	28 hours for infrequent users 56 hours for frequent users

When your legs get weaker, time starts running faster.
—Mikhail Turovsky, Russian-American artist and aphorist

How long would it take to walk (or run) off these indulgences?[4]

The food (calories)	Walking	Jogging
Cheeseburger (630)	2 hours, 55 minutes	1 hour, 12 minutes
Cinnabon (813)	3 hours, 45 minutes	1.5 hours
Caffe latte, whole milk (170)	38 minutes	16 minutes
White wine (120)	33 minutes	14 minutes
Steak, 4 oz (212)	1 hour	24 minutes
Bacon, 4 slices (207)	56 minutes	24 minutes
Bagel w/cream cheese (390)	1 hour, 48 minutes	45 minutes
Brownie (112)	31 minutes	13 minutes
Pizza, 1 slice (272)	1 hour, 15 minutes	31 minutes
McFlurry with Oreos (335)	1 hour, 33 minutes	38 minutes

*Based on a 35-year-old woman who is 5 feet, 7 inches tall and 144 pounds.

Time Trivia: How often do men and women cry—and for how long?

According to crying expert William H. Frey II, author of *Crying: The Mystery of Tears,* women cry, on average, 5.3 times per month, men 1.4 times. Other researchers have found lower frequencies, but the same gender difference: women consistently cry 2 to 5 times more than men.[5] Men tend to cry for between 2 and 4 minutes at a time, while women weep for 6 minutes. Crying turns into full-blown sobbing for women 65 percent of the time, but just 6 percent of the time for men. It's not clear why men cry less than women, but it might have something to do with testosterone.

> *All parts of the body which have a function,*
> *if used in moderation and exercised in labours in which*
> *each is accustomed, become thereby healthy, well-developed*
> *and age more slowly, but if unused and left idle they become*
> *liable to disease, defective in growth, and age quickly.*
> —Hippocrates, the Father of Medicine

Sleep and Dream Time

How many hours do we need to sleep?

In general, adults need 7 to 8 hours a night. Children and teens need more. There's good evidence that sleeping too much (over 9 hours) or too little (4 to 5 hours) can be bad for your health.

Studies have found that people who over- or undersleep have higher mortality rates than those who get 7 to 8 hours of snooze time.[6]

Age	Sleep need
Newborns (up to 2 months)	12 to 18 hours
Infants (3 to 11 months)	14 to 15
Toddlers (1 to 3 years)	12 to 14
Preschoolers (3 to 5 years)	11 to 13
School-agers (5 to 10 years)	10 to 11
Teens (10 to 17 years)	9 to 10
Adults (18 and up)	7 to 8

How many hours should teens sleep to do well in school?

Perhaps less than you think. A recent study upended current notions about how much sleep teens need in order to perform optimally. Researchers at Brigham Young University found that teens 16 to 18 years old did better on tests when they slept 7 hours each night, rather than the prescribed 9.

Age	Optimal hours of sleep for performance
10-year-olds	9 to 9.5
12-year-olds	8 to 8.5
16-year-olds	7

These findings seemed a bit odd, so I sent a note to Mark Showalter, one of the authors of the study. "If a 16-year-old sleeps a modest 8 hours a night, is he facing potentially lower scores in school?" I asked. It's not quite that definitive, he answered. "One way to read our results," Showalter explained, "is that 7 hours is optimal; but it wouldn't be far off to say that how much sleep you get *within* the normally observed range of 7 to 9 hours probably doesn't make much difference on performance. It's only when you get in the extreme tails that problems start popping up. And I suspect there is important individual variation that our data can't unmask." Which I took to mean, don't change your teen's sleeping habits just yet!

Time Trivia: Sleep dept

Most high school students have a sleep debt of 5 to 10 hours by the end of the school week.[7]

Mirror mirror on the wall, who is the most sleep deprived of us all?

A recent government study found something astounding: 30 percent of American adults sleep for 6 or fewer hours each day. Most doctors would say that is way too little. Who is the *most* sleep deprived? Those who work the night shift (44 percent sleep 6 or less hours), blacks (38.9 percent), and those who are widowed, divorced, or separated (36.4 percent).[8]

Time Trivia: How much sleep do your dog and your armadillo need?

Dog owners might be surprised to learn that their pets only need 2 more hours of sleep than they do. My dog seems to spend half the day curled up in his bed. Here are animal snooze times to ponder.[9,10]

Animal	Snooze time
Little brown bat	19.9 hours
Giant armadillo	18.1
Lion	13.5
Cat	12.5
House mouse	12.5
Dog	10.6
Red fox	9.8
Chimpanzee	9.7
Rabbit	8.4
Human	8.0
Cow	4.0
Horse	2.9
Giraffe	1.9

How many times do we yawn a day?

The average adult yawns five times a day.[11] You'll yawn more if you're very sleepy or if you are hanging around a yawner—yawning is contagious. Scientists don't know for certain why we yawn. One of the latest theories is that yawns help to cool the brain off. When the brain gets too warm, the body takes in cool air via a yawn to bring down the heat.[12]

How much time do we spend dreaming?

The standard answer is this: humans have about 4 to 5 dreams every night and each one lasts from 5 to 30 minutes. But some researchers believe we spend more like 70 percent of our sleeping hours dreaming or in a dreamlike state called "sleep mentation" (or non-REM dreams).

Although the word *dream* typically has positive connotations, our dreams are anything but "dreamy." By studying dream journals, researchers have learned that the majority of the dreams we have are bad ones. The most common emotion experienced while dreaming? Anxiety.

How much time do we spend napping?

On a typical day, 1 out of 3 Americans take a nap, according to a 2009 survey by the Pew Research Center. However, the survey did not ask how long those naps lasted. They could have been anywhere from 10 minutes to 2 hours.[13]

How long *should* a nap be? There's some debate. A study in the journal *Sleep* looked at the benefits of naps of various lengths and found that a 10-minute nap produced the most benefit in terms of reduced sleepiness and improved thinking.[14] However, other studies have found that longer naps of 40 to as much as 90 minutes can actually help consolidate memories and learning. The reason for the debate might be because individuals differ in their napping needs and napping reactions.

Time Exercise 4: Preparing for Sleep Deprivation

If you're planning on pulling a few all-nighters, stock up on sleep in advance. One study found that sleeping 10 hours a night, a week before a period of sleep-deprived nights, helped people recover their cognitive faculties more quickly than trying to make the sleep up on the other end—after the deprivation occurred. Sleeping 10 hours a night on demand, is not possible for many people. The point is, be as well rested as you can going into a high-stress, low-sleep period of time (like finals).

How long does it take to get over jet lag?

About one day per time zone crossed. But it also depends on which direction you are flying. Going east is harder on your body than going west. Eastward travel makes it difficult to fall asleep, and westward travel makes it difficult to stay asleep. Older people have a harder time adjusting than younger folks. Morning people are more affected than night owls. Here's one fact that's true for everyone: sunlight helps your body adjust to a new time zone, so when you're traveling, spend lots of time outside.

Time Spent in Medical Care

How long is the typical office visit?

The average doctor's visit today lasts about 22 minutes, up from 16 minutes in 1989. Yes, that's right: time spent with doctors has actually risen over the past two decades according to the National Center for Health Statistics. Psychiatrists spend the most time with their patients, an average of 33 minutes in 2006. Pediatricians spend the least, just 17 minutes.

A 2012 report by Medscape[15] found slightly lower numbers: half of doctors spend 16 minutes or less with patients, and half spend 17 minutes or more. A devoted 20 percent of doctors spend 25 minutes or more with their patients. Women generally spend more time with their patients than men. "Female physicians tend to concentrate in obstetrics and primary care, where doctors naturally spend more time with patients than do other specialists," the report explained. In addition, women are often "part-time employees for lifestyle reasons, and their main priority isn't on maximizing revenue for the practice. So, they are able to spend more time with patients." (This survey does not include psychiatrists.)

Time spent with each patient	Percentage of doctors
Less than 9 minutes	6
9 to 12 minutes	18
13 to 16 minutes	16
17 to 20 minutes	21
21 to 24 minutes	10
25 minutes or more	20

What's the average hospital stay?

Patients who check into the hospital stay for 4.8 days on average, down from 6.8 days in 1990. Length of stay depends on one's diagnosis, age, and sex. Older folks tend to stay in the hospital longer than younger ones; men generally have longer stays than women.[16]

How long does a drive-through hospital visit take?

Stanford Hospital conducted a study to determine whether treating moderately ill patients in their cars was a good idea. (Only in California!) The hospital found that for mild problems, like the flu, it was much more efficient for patients to stay in their cars than go inside. Doctors could treat influenza patients in a drive-through clinic in an average of 26 minutes, compared to 90 minutes when seen inside the hospital.[17]

When is the worst time to visit the hospital?

July. At teaching hospitals responsible for training new doctors, patient death rates increase in July, while efficiency in patient care decreases, according to a report published in 2011 in the *Annals of Internal Medicine*. The report, which reviewed 39 previous studies, found that death rates increased by 8 percent, to 34 percent in July.[18]

Doctor Time

How many hours do docs in the United States spend on billing?

While we're looking at how much time *we* spend in medical care, let's look at how much time medical professionals spend making it all possible. According to a study published in *Health Affairs*, nursing staffers, including medical assistants, spend 20.6 hours per physician per week dealing with health insurance plans. That's a lot of time, and it costs doctors a lot of money. The study's authors asked hundreds of doctors and administrators in private practices in the United States and Canada how much time they spent filling out insurance-related paperwork and dealing directly with insurers on billing matters. In Canada, which has a single-payer health system, medical offices spend one-tenth the time on billing that their U.S. counterparts do.[19]

The Medscape survey looked at how much time doctors alone spent on paperwork and found that the majority devoted just 1 to 4 hours a week to administrative activities. However, a small percentage of doctors (13%) spent more than 25 hours per week on paperwork! The doctors most apt to have a time-consuming paperwork load were HIV/AIDS physicians, pathologists, and oncologists.

Mind

How long does is it take for the brain to fully mature?

A woman's brain reaches full maturity at about 22 years, a man's at 25. (No comment!) But let's back up. Your brain reaches 95 percent of its adult size by the time you're 6. Once you hit adolescence, around age 11 in girls and 12.5 in boys, the gray matter in your brain is at its thickest. Gray matter is responsible for communication with other brain cells, among other functions. As you move through adolescence, the brain begins to remodel itself. The brain sheds connections that are underused, which causes your gray matter to become thinner but more efficient. While gray matter is thinning, white matter is thickening. White matter, or myelin, protects the never fibers and speeds up their transmission. So, as you move into early adulthood, you have fewer but faster connections. Think of it this way: a teenager has more possibilities, but an adult has more focused expertise.[20]

How quickly can the brain retrieve a memory?

The brain can be speedy when the task is simple. To retrieve a memory that is fairly basic, it takes the brain a mere .6 to .8 seconds. When you're older, or when a fact is less familiar, it might take your brain minutes, even hours to retrieve a piece of information. (Quick: name the titles of the last three books you read.)

The human brain's storage capacity is more than 1 million gigabytes. If your brain were a DVR, it could hold 3 million hours of TV shows, according to Paul Reber, a psychologist and professor at Northwestern University.

How many words can a human understand in a minute?

Audiobooks are typically spoken at a leisurely pace of 150 to 160 words per minute. Fast-talking auctioneers and radio announcers can speak at a clip of 350 to 400 words per minute, but that's the outer limit of a human's ability to comprehend language. The former Finnish radio and TV announcer Raimo Häyrinen could speak at a pace of 471 words per minute!

Blind people, it turns out, have superhuman listening powers. Children who become blind between the ages of 2 and 15 can comprehend speech that is sped up to 25 syllables per second, or approximately 750 words per minute (a rate that no one really speaks at, but that researchers can create via a synthesizer), according to a study conducted at the Hertie Institute for Clinical Brain Research in Germany.

Why is this? The researchers examined the brains of blind and sighted people using magnetic resonance imaging and found that the blind subjects were using a part of the brain that normally responds to vision to comprehend speech. The blind folks' brains had essentially taken over the visual cortex for auditory purposes, which helped to supercharge their listening power. How smart are our brains? Very.

Time Trivia: Memorization champ

Simon Reinhard, a German memory athlete and lawyer, can memorize the order of a randomly organized deck of cards in 21.19 seconds. He has held the world record for this feat since 2010.

How much time do Americans spend in "serious psychological distress"?

This is a question the government has been asking its citizens every two years since 1997. On average, Americans spend 3.2 days a month in serious distress. Some details:[21]

- Women have more bad days (3.6) than men (2.7).

- Lower-income folks have more distress than higher-income folks.

- Whites, blacks, and Hispanics have about the same amount of distressed days. Asians have considerably fewer bad days than other races.

- Distress seems to peak in middle age and then decline steadily. People between 45 and 54 have the greatest number of bad days; people between 65 and 74 have the fewest. This research squares with the findings from a recent AARP study, which found that people are happiest in their 60s and 70s. See Chapter 1, Happiest years of life, page 31.

The numbers have stayed relatively stable over the past decade.

How much time do we spend daydreaming?

A lot. Nearly half our waking hours are spent disconnected from the present moment. Two Harvard psychologists, Daniel Gilbert and Matthew Killingsworth, used an iPhone app to keep in touch with 2,250 volunteers. The app contacted the users at random intervals, asking whether they were thinking about the activity they were engaged in or whether they were thinking about something else and if so, whether that "something else" was pleasant, neutral, or unpleasant. The app also tracked how happy the folks were and what they were currently doing.

The study found that the subjects' minds were wandering 46.9 percent of the time. That's a lot of daydreaming. But more startling was the authors' finding that daydreaming was correlated with unhappiness. "People were less happy when their minds were wandering than when they were not, and this was true during all activities, including the least enjoyable," the authors wrote. "Although negative moods are known to cause mind wandering, time-lag analyses strongly suggested that mind wandering in our sample was generally the cause and not the consequence of unhappiness." Think of it this way: when your mind is wandering, you're usually worrying about something from the past or something in the future—rather than simply enjoying and soaking up the present moment.[22]

I think some people from the West, where technology is so good, think that everything is automatic. You should not expect spiritual transformation to take place within a short period; that is impossible. Keep it in mind and make a constant effort, then after 1 year, 5 years, 10 years, 15 years, you will eventually find some change.
—The Dalai Lama, *The Book of Wisdom*

How much time do we spend meditating?

In 2007, about 20 million Americans (9.4%) reported that they had meditated over the past 12 months. Considering how much time our minds spend wandering and how much unhappiness it brings (see above), it's understandable and laudable that people would be attracted to a discipline that helps the mind stay alert to the here and now.

For how long *should* you meditate?

There are no *should*s when it comes to meditation, *but* it appears that 30 minutes a day is the amount of daily meditation time needed to change your mind. A study published in 2011 in the journal *Neuroimaging* found that individuals who meditated for 30 minutes a day over an 8-week period had increases in gray matter in "regions involved in learning and memory processes, emotion regulation, self-referential processing, and perspective taking." The study participants practiced mindfulness-based stress reduction, a form of meditation in which practitioners focus on one thing, like their breath or a part of their body. Other studies have found that regular meditation can lower blood pressure, increase attention spans, and activate the area of the brain associated with good feelings.[23] But even 10 to 20 minutes a day can have a profound effect on your mental outlook. The key is consistency.

What's the best time to make a decision?

When you don't have lots of decisions to make. Or just after you've had a short break and a light snack. One study examined factors that influenced judicial decisions. Researchers analyzed 1,112 rulings by Israeli judges about whether prisoners should be granted parole. When judges had to make many rulings in a row, without a break, they tended to rule in favor of the status quo, meaning keeping the prisoner in prison. If the judges took a break and had a bite to eat, though, their mental resources were restored and they were more likely to grant parole. What do legal decisions have to do with other decisions? The same distractions can apply to other decision-making situations, particularly if you have to make many decisions in a row—like at work, or on vacation, or at your lawyer's office.[24]

 Book Excerpt: How long does it take to reach enlightenment?

A young man in Japan arranged his circumstances so that he was able to travel to a distant island to study Zen with a certain Master for a three-year period. At the end of the three years, feeling no sense of accomplishment, he presented himself to the Master and announced his departure. The Master said, "You've been here three years. Why don't you stay three months more?" The student agreed, but at the end of the three months he still felt that he had made no advance. When he told the Master again that he was leaving, the Master said, "Look now, you've been here three years and three months. Stay three weeks longer." The student did, but with no success. When he told the Master that absolutely nothing had happened, the Master said, "You've been here three years, three months, and three weeks. Stay three more days, and if, at the end of that time, you have not attained enlightenment, commit suicide." Toward the end of the second day, the student was enlightened.

Occupation
Education & Work

What do school and work have in common? Both are compulsory and both are time-consuming. Most of us *have* to attend school, and most of us *have* to hold down a job. The hours add up. Some of those hours are satisfying (learning, socializing, making money), and some are not so satisfying (commuting, homework, meetings).

Let the information in this chapter inspire you to take action—in the college you select, in the vacation time you use, in the commute you choose, and in the meetings you arrange. Life is more enjoyable when you feel in control of your time.

Education

Education has become increasingly important the world over. More education typically translates into more opportunities and income. In the United States, increasing numbers of teens go to college and more graduate with degrees. Today, 30 percent of American adults have a bachelor's degree, up from 11 percent who had BAs in 1970.[1] Pundits predict that college enrollment will set new records through fall 2020.[2] No wonder, then, that high school grads have such a difficult time getting into the college of their dreams; more kids equals more competition. In addition, women are becoming the majority; they now account for 57 percent of college students and 60 percent of graduate students.

School Hours

How many hours do kids spend in school per year?

It depends on the state. The US government doesn't dictate the number of days that kids must go to school each year, nor does it dictate how long the school day must be. Each state decides school hours and year length for itself. On average, the school year is 180 days, the school day is 6.6 hours, and the total annual school hours are 1,193. There's some variation between the states; consider these extremes.[3]

The award for *most instruction hours* goes to . . . **Texas!** Students in the Lone Star state spend 1,289 hours in school each year (180 days/year x 7.2 hours per day).

The award for *least instruction hours* goes to . . . **Minnesota!**
Students in the Land of Lakes spend just 1,102 hours in school
each year (176 days/year x 6.3 hours per day).

How much time did kids spend in school during Reconstruction?

We've come a long way. Back in 1869, as the country was recovering from the Civil War, the school term was 132 days. School was not necessarily a family's or a child's first priority, with just 59 percent of students attending on a daily basis. Today, the average school year is 180 days, and 93 percent of kids attend regularly. Let's zoom back in time.[4]

	1869–70	**1909–10**	**1949–50**	**2008–09**
Average length of school term	132 days	158	178	180
Average number of days students attended school/year	78	113	158	167
Percentage of enrolled students who attended school daily	59	72	89	93

How do US school hours compare to other countries? (Note to kids: Don't move to Indonesia.)

Middle-school students in the United States spend more time in school each year than the global average of 1,011 hours. But the formula for annual school hours varies dramatically from country to country. In some places, like Thailand, kids have long days but short school years, while in others, like Armenia, it's the opposite. Here's how school time for eighth graders compares in 10 select countries across the globe, including the highest- and lowest-ranking countries in the 49-country sample.[5]

Instruction time for eighth graders around the globe

Country	Instruction hours/year	Instruction days/year	Instruction hours/day
Indonesia	1,385	231	5.6
Thailand	1,264	166	6.2
Chinese Taipei	1,193	202	5.6
United States	1,109	180	5.8
Armenia	1,097	210	4.7
Japan	1,016	201	4.9
Average	1,011	190	5
Korea	990	206	4.4
Sweden	942	179	4.9
Russian Federation	867	199	4
Syria	799	166	4.4

How do schoolkids in the United States spend their day?

Students in first through fourth grade receive 31 hours of instruction each week, and 22 of those hours are spent on four main subjects:[6]

Subject	Hours/week
English	11.7
Math	5.6
Social Studies	2.3
Science	2.3

How many hours do fourth graders spend in science class?

Where do the hours go? Fourth graders spent 89 hours in science class in 2007, well above the international average of 67 hours, and yet US students are ranked as *average* in their knowledge of science. Several countries with almost comparable economies ranked well above average in science smarts, including China, Hong Kong, Korea, Finland, Singapore, Japan, and Germany.[7] Ouch.

How many hours do eighth graders spend in math class?

Same question here. American students spend more than an average number of hours in math class, but rank *below average* in math knowledge.[8] Eighth graders received 148 hours of yearly math instruction (13% of total instruction time) in 2007—well above the international average of 120 hours. China topped the chart by giving eighth graders 158 hours of math instruction. And yet, when you compare kids by math *skills,* students in the United States are slightly below average in their knowledge of basic math concepts.[9] Ouch again.

Global education time

Kids in the United States spend a minimum of 13 years in school. That's not the norm in the rest of the world. Kids from poorer countries spend much less time in school than those from wealthy ones.[10] That's too bad, because education helps propel economic expansion. In the developing world, for instance, an extra year of primary school increases a girl's eventual wages by 10 to 20 percent. An extra year of secondary school increases her wages 15 to 25 percent, according to The Girl Effect, a nonprofit organization dedicated to improving the lives of girls. Economic development also slows the birthrate: the more education a woman has, the less likely she is to have children.

Annual Income	Years of education
Low income areas: $995 or less per year	7.9 years
Lower Middle $996 to $3,945	10.3
Upper middle $3,946 to $12,195	13.8
High $12,196 or more	14.5

Homework Hours

How much time do students spend on homework?

Ah, the homework wars. Over the past few decades, grade-school kids have been spending more and more time on homework. Meanwhile, college kids have been spending less and less time on *their* studies. Consider the following numbers (the most current information available):

6- to 8-year-olds: In 1981, this age group spent 52 minutes a week on homework; in 2003, that number had climbed to 2 hours and 36 minutes.

9- to 12-year-olds: In 1981, these kids spent 2 hours and 50 minutes a week on homework; in 2003, they spent 4 hours and 20 minutes.[11]

College kids: In 1961, college kids spent about 24 hours a week on schoolwork; in 2003, they spent a mere 14 hours, according to a study by economists Philip Babcock and Mindy Marks. The economists found that study time fell for students from all demographic groups, for students who worked and for those who did not, for students within every major, and at four-year colleges of every type, degree structure, and level of selectivity. Why the drop off? The authors theorize that students today are forgoing study time for leisure time and that academic standards are falling at institutions of higher education.[12]

Does homework help?

Maybe. Maybe not. In a recent survey, the majority of college presidents (58%) said that public high school students arrive at college less well prepared than students of a decade ago[13]; half

(52%) of college presidents agree with Babcock and Marks's findings that college students today study less than their predecessors did a decade ago; only 7 percent say they study more.[14]

 Time Trivia: Homework booster

High school students who have a parent with a college degree are more likely to do homework (and devote more time to their studies) than students whose parents have less education. On an average day, 54 percent of teens who have a parent with a BA degree or higher do homework, compared to just 39 percent of teens with a parent who had not completed college.[15]

Completing College

How long does it take to complete a college degree?

For-ever! Back in the 1970s and 1980s, students spent four years on their undergraduate degrees. Today it takes students more than six years to complete a "four-year degree."[16] In 2008, just 58 percent of college students were able to graduate in 6 years or less.[17] Many students are not finishing their degrees at all. The United States has the highest college dropout rate in the industrialized world.[18]

Students working on master's degrees take about 7.2 years to complete their degrees; those working on doctorates spend more than 9 years getting their degree.

Why does it take so long to finish college?

Cost is one reason. Between 1980 and 2010, the price tag for a four-year private education nearly tripled; the price for a four-year public education (where the majority of students are) *more* than tripled.[19] Students who obtain a four-year degree are saddled with, on average, $23,000 worth of debt. Some students drop out because they can't afford the tuition. Another culprit is busy schedules—75 percent of students commute to college in order to juggle multiple commitments. Poor management is yet another reason. Students aiming for a bachelor's degree should take 120 credits, but on average those students take 136.5; students working toward an associate's degree also overload on credits, taking 79 when they only need 60, according to Complete College America. Lastly, cash strapped public colleges are often unable to offer enough of the required courses kids need to finish on time.[20]

Work

The amount of time the average American spends working hasn't changed appreciably over the past few decades.[21] But most workers feel more burdened: they are expected to work harder, and technology makes the workday feel longer.

The workforce, though, is slightly smaller. Today, roughly 75 percent of all adults have some kind of paying job, down from 79 percent who did in 1973. Why the decline? There are multiple reasons, including the fact that more adults are now in school and more young adults are having difficulty finding paying jobs.

Work Time

Work hours by age and status

Here's a quick summary: Married men work *more* than unmarried men. Married women work *less* than unmarried women. Those with higher incomes and net worth spend more time working, regardless of the type of work they do, according to a study by the Sloan Center on Aging & Work at Boston College.[22]

Hours of paid work/week

	Under age 50 hours/week	50 to 64 hours/week	65 and older hours/week
Married men	42	36	8
Unmarried men	37	30	8
Married women	26	26	7
Unmarried women	32	30	3

How many hours do people work by profession?

The researchers at Pace Productivity tracked the number of hours 12 different professionals logged each week as well as the time they devoted to specific tasks during the day. Workers used a TimeCorder device to track their hours. (The weekly hours include business travel and breaks during the day plus hours worked at home and at a client's office; commuting, though, is not included.)[23]

Job description	Weekly work hours	Time spent on specific tasks (in minutes)
President/VP	59.7	32
Field supervisor	54	33
Consultant	54	35
Middle manager	50.6	24
University faculty	48.8	38
Sales manager	48.7	22
Outside sales rep	48.5	17
Retail bank sales	45.5	19
Clerical	45	14
Receptionist/assistant	44.6	10
Inside sales	43.6	6
Municipal worker	42.5	11

Who logs the most hours?

Professional men are the most likely to put in 50-plus-hour work-weeks. Not surprising, right? But women are catching up. The percentage of professional men and women who work extralong hours has been drifting upward for decades (see table below).[24]

**Percentage of men and women who work
50-plus hours a week: 1977 vs. 2008**

	Men		Women	
	1977	2008	1977	2008
Professional	34.0%	37.9%	6.1%	14.4%
Middle income	21.2	22.9	3.4	8.3
Low income	16.1	8.7	3.7	3.9

Business owners work longer hours than employees

It's a great fantasy, but a tough reality: running your own business takes time and lots of it. About 63 percent of business owners report working 40 or more hours per week. Compare that to the average worker who puts in just 34.4 hours. The hours are slightly less grueling if you don't have employees. Just 34 percent of owners of firms without payrolls worked 40+ per week, according to the 2007 Survey of Business Owners.

Stay-at-home workers put in longer hours, too

About 9 percent of people who worked exclusively at home in 2010 reported working 11 or more hours in a typical day. By contrast, just 7 percent of people who worked in an office put in 11+ hour days, according to the US Census.

Who are the busiest workers in the world?

The Mexicans! Why? In part because one-third of the Mexican workforce is self-employed-and self-employed folk tend to work longer hours and take less vacation time than those with full-time jobs. Economists also theorize that because the country is underdeveloped and inefficient, workers put in longer hours than their counterparts in the developed world in an effort to keep up. The Organisation for Economic Co-operation and Development (OECD) collects data from its member countries each year on working hours, among many other stats. The rankings change slightly from year to year.

OECD Country	Average annual hours actually worked per worker: 2011[25]	Percentage of workers who work 50+ hours per week[26]
Mexico	2250	29
Korea	2,090	22
Chile	2,047	7
Greece	2,032	5
Russian Federation	1,981	.2
Hungary	1,980	3
Poland	1,937	7
Estonia	1,924	4
Israel	1,890	19

Turkey	1,877	43
Slovak Republic	1,793	6
United States	1,787	11
OECD-Total	**1,776**	**NA**
Czech Republic	1,774	9
Italy	1,774	5
New Zealand	1,762	13
Iceland	1,732	11
Japan	1,728	30
Portugal	1,711	5
Canada	1,702	4
Australia	1,693	14
Spain	1,690	7
Finland	1,684	4
Sweden	1,644	1
Switzerland	1,632 (2010)	6
United Kingdom	1,625	12
Luxembourg	1,601	4
Austria	1,600	9

Belguim	1,577	4
Ireland	1,543	4
Denmark	1,522	2
France	1476	9
Norway	1,426	3
Germany	1,413	5
West Germany	1,399	NA
Netherlands	1,379	.7

Long hours = bad heart

Dial back, workaholics: long hours are linked to heart disease. A study published in the *Annals of Internal Medicine* found that Brits who regularly worked 11-hour days or longer were 67 percent more likely to develop heart disease than those who worked 7- or 8-hour days. Do long work hours *cause* heart disease, or do people who work long hours have other bad habits, like poor diets and fewer hours of exercise, that might contribute to heart problems? It's impossible to say for sure, but if you put in long days on the job, you might want to get your heart checked.[27]

*One cannot become a saint
when one works sixteen hours a day.*
—Jean-Paul Sartre

Who invented the 40-hour workweek?

Unions pushed for shorter workweeks in the early nineteenth century, but the 40-hour week didn't catch on in the United States until the dawn of the twentieth century. Henry Ford helped. In 1914, Ford doubled workers' pay and cut their shifts from 9 hours to 8. George F. Johnson—whose Endicott-Johnson shoe factories made all the military boots for US soldiers—also contributed to the movement. In 1916, Johnson announced that his plants would adhere to a 40-hour workweek. (Johnson implemented other progressive policies, too: his workers received above average pay and comprehensive medical care.) Why were these guys so kind? Because they cared about output. In the early 1900s, managers found that cutting work hours greatly improved worker productivity. Tell that to your boss!

 Pop Quiz: Work Hour Rules

Q: What is the limit on the hours an employee can work each day or week?

A: There is no limit. That's right, *no limit.* According to the federal Fair Labor and Standards Act (FLSA), there is no upper limit on the number of hours per day or per week that employees aged 16 years and older can be required to work.

Q: How many breaks must employees be given each day?

A: Zero. The FLSA does not require that employers give breaks to their workers. However, some states, such as California and Washington, have enacted their own laws on this matter. According to the FLSA: "If you work in a state which does not require breaks or meal periods, these benefits are a matter of agreement between the employer and the employee (or the employee's representative)."

Q: How much notice must an employer give an employee before letting him go?

A: None. The FLSA does not require employers to give notice to their workers prior to a termination or layoff. Some states, however, have their own notification policies.

What's the average job tenure in America?

In 2011, workers had been at their posts a median of 4.6 years. Workers between the ages of 25 and 34 had been at their current job for 3.2 years, while workers 55 to 64 had been on the job for 10 years. Folks in the public sector had nearly double the tenure of those in the private sector, 7.8 versus 4.2 years. Overall, 34 percent of workers had been on the job for more than 10 years.[28]

For how many years will *you* work?

Chances are high that you'll be living *and* working longer than your parents. To cope with longer life spans and dwindling benefits, Americans are delaying retirement. The average adult now plans to retire at age 67, up from age 63 in 2002 and age 60 in the 1990s. Adults hope to plump up their savings by tacking on a few more years of work before officially calling it quits.

Times are tough. Just 38 percent of Americans say they will have enough money to live comfortably in retirement, way down from 59 percent who felt that way in 2002.[29] What's more, 75 percent of middle-class workers expect to work during their retirement. And, *yikes*, 25 percent of workers say they will need to work until they are 80, according to a 2011 retirement survey by Wells Fargo.[30]

Time Spent Looking for Work

How long does it take to find a job?

Once you leave you a job, how long does it take to find a new one? When the economy is healthy, it takes just 10 weeks or so; when it's not healthy, much longer. For instance, during the 2008–2010 economic downturn, it took workers 6 months on average to find new positions—the longest job-hunt time on record. The previous high, in May 1983, was 12.3 weeks, less than half the level of 2010.

Long stretches of unemployment can damage more than your pocketbook. According to one survey, 38 percent of those who were unemployed for 6 months or more admitted they lost some self-respect, compared with just 29 percent of the jobless who were out of work for shorter stints. The long-term unemployed were more likely to seek professional help for depression or other emotional issues (24% vs. 10% for those unemployed less than three months).[31]

How much time do job seekers spend online?

The job market has been up and down over the past few years, and job seekers have been spending more time online checking out career and job sites. Here are the sites that got the most visitors in July 2012, plus the average time those visitors spent using them.[32]

Site	Minutes per Visitor	Total Unique Visitors
Indeed	13.8	17,965
CareerBuilder LLC	10.7	21,618
Monster, Inc.	9.9	22,375
BrassRing.com	9.4	4,790
SimplyHired, Inc.	5.2	5,529

On-The-Job Time

How much time do workers waste?

Back in 2008, when surveyors at Salary.com asked this question, 73 percent of workers said they spent part of their day on nonwork activities. How much time exactly did they waste? A brazen 14 percent of workers reported wasting 3 hours or more a day; 22 percent wasted 2 hours; 64 percent wasted one hour or less. What did they spend their time on? Surfing the Net, socializing with colleagues, and doing personal business were the top three time wasters.

Why do workers goof off? The workers in the survey reported that they felt unsatisfied, underpaid, or didn't have deadlines or incentives. This survey was done during the recession, so the results may have been influenced by the poor economy. (Unfortunately, Salary.com was bought by Kenexa and stopped doing this survey.) Workers probably goof off even more these days, because the distractions are even more alluring: Facebook, Instagram, texting, shopping on Gilt—just to name a few.[33]

What cuts into worker productivity?

Workers who participated in the Salary.com survey fingered these time sappers:[34]

1. Fixing someone else's work

2. Dealing with office politics

3. Waiting for a worker to finish something you need

4. Attending work-related meetings or events

5. Administrative work

Does wasting time *waste* time?

That's not a trick question. Researchers have found that certain types of goofing off can make workers *more* productive. A recent report, "Cyberloafing at the Workplace: Gain or Drain on Work?," found that web surfing serves an important restorative function and helps reenergize the mind after an intense bout of work. Consider this finding: A group of participants spent 20 minutes on a work task highlighting as many "e's" as they could find in a piece of text. The individuals were then split into three groups. One group could do what ever it liked for 10 minutes, another did a simple task, and the third surfed the web. After their 10-minute break, the participants circled letters for 10 more minutes. Guess what? The group that had surfed the web was more effective at the circling task than those in the other two groups. What's more, the surfers reported lower levels of mental exhaustion and boredom and significantly higher levels of engagement than the two other groups.

Surfing has a restorative function, the researchers found, but e-mail does not. E-mail, in fact, has a negative effect on productivity and mood. (No surprise there!) The authors concluded that Internet browsing enhances psychological engagement with work—and possibly job creativity as well—whereas reading e-mail negatively affects employees' ability to concentrate on work.

Cyberloafing, like yoga for the mind.

 Time Exercise 5: Create your own 1-minute pitch

Harvard Business School's online tool, *HBS Elevator Pitch Builder*, helps workers create a 1-minute pitch about themselves or their business. The average pitch created on the site lasts an average of 56 seconds and has 231 words (as of 2012).

How often are workers interrupted?

About every 3 minutes. Researchers shadowed a group of workers for three and a half days and timed their activities down to the second. When the worker picked up the phone, the researchers started the clock; when the worker put it down, the researchers stopped it. The study, by Gloria Mark, professor of Informatics at UC Irvine, found that workers switched between small tasks such as phone calls, documents files, and Internet surfing, every 3 minutes and 5 seconds, on average. Most interesting: almost half (44%) of these were self-interruptions—meaning they weren't *answering* the phone; they were *picking* it up to initiate a call themselves. The rest, 56 percent, were external interruptions from pesky coworkers or a ringing phone.[35]

How long does it take to get back on task?

Twenty-three minutes. According to Mark's research, 82 percent of all interrupted work is resumed on the same day. But once a worker gets interrupted, it takes 23 minutes and 15 seconds to get back on track.

Time Trivia: How much time do we spend on e-mail?

Office workers spend an average of 2.6 hours each day dealing with e-mail messages, that's about 28 percent of the work day! They spend another 1.5 hours looking for information or tracking down colleagues for help with a specific task, according to a survey by McKinsey Global Institute.

When is the best time to schedule a meeting?

Tuesday at 3 P.M., according to the online scheduling service Whenisgood.net. The service studied 100,000 responses to 34,000 meeting requests (primarily from customers in the United Kingdom, Canada, and the United States) and found that Tuesday at 3 P.M. was the optimal meeting time. The study also found that workers are most flexible between 10 A.M. and 11 A.M. and that, on average, only 3 to 4 people out of 10 will be available at any given time to attend a meeting. One surprise: "People are happier than you might think to work through lunch," the study found. Of course! That's because hardly anyone takes a real lunch these days. See "How much lunchtime do workers actually take?" on page 174.[36]

Meeting efficiency tips

"If you had to identify, in one word, the reason why the human race has not achieved, and never will achieve, its full potential, that word would be 'meetings,'" wrote the humorist Dave Barry.[37] Does anyone love spending time in meetings? Steve Jobs did, but most workers don't. If you regularly schedule meetings, here are ways to make them more efficient, courtesy of Pace Productivity.

- Write an agenda and distribute it before the get-together. It should list the purpose of the meeting and the items to be discussed; the list of items should be specific and focused. Include a hard stop, too. End times can bring a needed sense of urgency to a meeting.

- Start meetings on the half hour. Research indicates that meetings are more likely to start on time when they are scheduled on the half hour, rather than on the hour.

- Schedule meetings in the afternoon. Time studies show that meetings are shorter later in the day. Workers become more efficient as quitting time approaches.

Time Off

How much paid vacation time do workers receive?

American workers get, on average, 25 paid days off each year (15 vacation days plus 10 holidays). But, unlike most countries, those 25 days are at the discretion of the employer. The United States is one of the only countries in the world that *does not require paid time off* for workers. Go ahead and envy workers in Finland, Brazil, and France, who are entitled to 40 to 41 paid days off a year (including public holidays). But *guaranteed* time off does not always translate into *more* time off. Pity workers in Canada who get just 19 days off, including holidays, and Chinese workers who get just 21 days, according to a survey by Mercer Consulting.[38]

How much paid time off *should* workers receive?

In 2012, Switzerland rejected a motion to give its workers 6 weeks of paid vacation, keeping the quota at 4. What would Americans vote for—if they actually had a say? Fully 44 percent of workers think American workers should have 4 weeks of paid vacation each year; 19 percent think workers should have 6 weeks; and 16 percent think workers should have 8, according to a *60 Minutes/ Vanity Fair* poll.

How much of their paid vacation days do workers actually *take*?

Workers may complain about their paltry vacation allotment, but the truth is just 57 percent of American workers take all their vacation time. A global survey of workers found that French employees were the most likely to take all their vacation time (89%) and that Japanese workers were the least likely (33%) to take advantage of paid time off.

There are clear regional differences, but income does *not* appear to play a role in the vacation-taking habits of the 12,291 workers surveyed in 24 countries. The survey authors note that among high- and low-income earners, the rate is the same: "about two-thirds take all their vacation time."[39]

Country	Percentage who take all their vacation time
France	89
Argentina	80
Hungary	78
Britain	77
Spain	77
Saudi Arabia	76
Germany	75
Belgium	74

Turkey	74
Indonesia	70
Mexico	67
Russia	67
Italy	66
Poland	66
China	65
Sweden	63
Brazil	59
India	59
Canada	58
United States	57
South Korea	53
Australia	47
South Africa	47
Japan	33
Overall Average	**65**

How long does it take to feel relaxed on vacation?

Employees say that it takes them 3 days, on average, to relax and unwind when they take a break from the office, according to the Families and Work Institute's 2004 study, *Overwork in America*. The study also found that "longer vacations, of 7 days or more, are associated with better psychological outcomes than shorter vacations."

How much lunchtime do workers actually take?

Surprise, surprise: a worldwide poll found that just 30 percent of American workers take a full lunch break, compared to 58 percent of French workers who do. After French employees, Italian and Indian workers were the most likely to take a full lunch break (48% do). Spaniards were the least likely to take lunch. Whatever happened to the siesta?[40]

Commuting Time

Guess who spends the most time getting to work in the United States?

Marylanders! The Americans with the highest incomes also have the longest commutes to work, 32.2 minutes. New Yorkers follow close behind at 31.5. New Jerseyans come in third, at 30.5 minutes.[41]

State	Mean Travel Time to Work
Maryland	32.2
New York	31.5
New Jersey	30.5
District of Columbia	30.1
Illinois	28.2
Massachusetts	28.0
Virginia	27.7
California	27.1
Georgia	27.1
New Hampshire	26.9
Pennsylvania	25.9
Florida	25.8
Hawaii	25.7
West Virginia	25.6
United States	25.5
Washington	25.5
Delaware	25.3

State	Mean Travel Time to Work
Connecticut	25.0
Arizona	24.8
Texas	24.8
Colorado	24.5
Louisiana	24.5
Tennessee	24.2
Michigan	24.1
Nevada	24.1
Alabama	23.9
Mississippi	23.9
South Carolina	23.6
Indiana	23.5
Maine	23.4
North Carolina	23.4
Rhode Island	23.4
Missouri	23.1
Ohio	23.1

Minnesota	23.0
Kentucky	22.9
Oregan	22.5
Vermont	21.9
Wisconsin	21.9
Utah	21.6
New Mexico	21.4
Arkansas	21.3
Oklahoma	21.1
Idaho	19.7
Kansas	18.9
Iowa	18.8
Alaska	18.4
Montana	18.2
Nebraska	18.1
Wyoming	18.1
North Dakota	16.9
South Dakota	16.9

Time Trivia: Commuter Facts

- The most congested hour of the week is Friday from 5 to 6 P.M.

- Morning congestion peaks between 7:45 and 8:00 A.M.

- Evening congestion peaks between 5:30 and 5:45 P.M. Monday through Wednesday and between 5:15 P.M. and 5:30 P.M. on Thursday and Friday.[42]

- 3.2 million Americans have extreme commutes of 90 minutes or more a day.

- 16.5 million commuters (12.4%) leave for work between midnight and 6 A.M.

Which international city has the worst traffic snarls?

Moscow. A survey of commuters in 20 cities around the world found that drivers in Moscow experienced the greatest delays: 45 percent of drivers reported they had been stuck in traffic for 3 hours or more. Nairobi was a close second; 35 percent of drivers reported 3-hour delays. Commuters in Mexico City, Beijing, and Shenzen routinely experienced delays of about 2 hours, according to IBM's 2011 Global Commuter Pain Survey.[43]

Longer commutes are linked to lower well-being

Americans who spend a lot of time schlepping to and from work report more physical and emotional distress than those with shorter commutes, according to the Gallup-Healthways Well-Being Index.[44] Time to talk to your boss about working from home.

Minutes from home to work	Percentage who had recurrent back or neck pain in last 12 months	Percentage who are obese	Percentage who worried much of the previous day	Percentage who experienced enjoyment much of the previous day
0 to 10	24	24	28	88
11 to 20	25	25	29	87
21 to 30	26	26	31	85
31 to 45	27	26	32	84
46 to 60	29	26	32	83
61 to 90	30	28	34	82
91 to 120	33	30	40	80

Your work is to discover your work and then with all your heart to give yourself to it.
—Buddha

Creation
Art, Music, Film & Architecture

How long did it take to build the Empire State Building, paint the Sistine Chapel, write *Gone with the Wind*? When you consider many great creations, time is part of the story. Mozart was famous for composing very quickly, whereas Beethoven had the opposite reputation (slow and deliberate but no less impressive). Picasso, enraged about the Spanish Civil War, painted his famous war mural, *Guernica,* in just over a month.

Deadlines, in fact, can spark creativity. National Novel Writing Month (NaNoWriMo) is wildly popular as it challenges writers to create a 50,000-word novel in a month. Likewise, 24-Hour Comics Day invites writers and artists to create a 24-page comic in 24 consecutive hours. Although time and creativity may seem unlikely companions, they are often inextricably, sometimes even fabulously, linked.

Art

How long *did* it take for Michelangelo to paint the ceiling of the Sistine Chapel?

About 4 years. Michelangelo di Lodovico Buonarroti Simoni worked on the ceiling from 1508 to 1512, at the commission of Pope Julius II. Not a long time, when you consider that he painted nine scenes from the Book of Genesis, including the iconic *Creation of Adam*, which included 343 figures. It was tiring work and, contrary to popular myth, he did not paint lying on his back. He stood on scaffolding, his neck craned upward. The smell and heat were intense, which prompted him to write a sonnet about the heinous conditions. Here's the beginning of Sonnet 5:

> *A goiter it seems I got from this backward craning*
> *like the cats get there in Lombardy, or wherever*
> *—bad water, they say, from lapping their fetid river.*
> *My belly, tugged under my chin, 's all out of whack.*
> *Beard points like a finger at heaven. Near the back*
> *of my neck, skull scrapes where a hunchback's lump would be.*
> *I'm pigeon-breasted, a harpy! Face dribbled—see?—*
> *like a Byzantine floor, mosaic. From all this straining*
> *my guts and my hambones tangle, pretty near.*
> *Thank God I can swivel my butt about for ballast.*
> *Feet are out of sight; they just scuffle around, erratic.*[1]

How long did it take for Leonardo da Vinci to paint *The Last Supper*?

Da Vinci worked on *The Last Supper* intermittently for 3 years, starting in 1495 and ending in 1498. The large painting, 15 feet × 29 feet, depicts Jesus explaining that 1 of his 12 disciples will betray him before sunrise. Lodovico Sforza, the Duke of Milan, commissioned the mural and the original is in the dining hall in the Convent of Santa Maria delle Grazie in Milan, Italy. The speed with which he created the mural is remarkable, considering Da Vinci was purported to be a terrible procrastinator (*Mona Lisa* took 20 years to complete). On his deathbed, Leonardo was said to exclaim, "Tell me, tell me if anything ever got done."

How long did it take Picasso to complete *Guernica*?

Just over 1 month. Picasso began sketching the initial design on May 1, 1937, less than a week after the Basque town Guernica was attacked by German and Italian planes during the Spanish Civil War. On May 11, he began to draw on the canvas. By early June, Picasso delivered the completed painting, more than 11 feet high and 26 feet wide, to the Spanish Pavilion of the Paris Exposition for exhibition.[2]

How long do people spend looking at art?

Typically less than a minute. A study conducted at the Metropolitan Museum of Art found that people looked at masterpieces from the museum's collection for an average of 27.2 seconds each.[3] And that's a relatively long amount of time. Other studies have found that museum visitors look at art for a few seconds before moving on. But remember, these are averages. Some visitors may spend minutes, perhaps even an hour, gazing at a piece of art and others a mere nanosecond.

How much time do people spend at museums?

The average museumgoer spends 2.4 hours on each visit. The peak attendance time at museums is between noon and 1 P.M. according to a study by the National Endowment for the Arts.[4]

How much time do American spend making art?

On any given day, 2.6 million people do some kind of creative activity, such as painting, sculpture, photography, or creative writing. Those who do these activities typically spend about 2.5 hours on them at a stretch.[5]

When do creative folks tend to do their work?

The peak time at which Americans paint, or sew, or do other artistic endeavors is between 2 and 3 P.M. and then again from 5 to 6 P.M.. Just as the craft people are calling it quits, writers are perking up. The percentage of people writing rises significantly at about 9 P.M. and peaks at 11 P.M.[6]

Writing

How long did it take Margaret Mitchell to complete *Gone with the Wind*?

Ms. Mitchell began working on her Civil War novel in 1927 while recuperating at home from a car crash. But she didn't turn in the final revision to her publisher until January 22, 1936. Why the lag? Ms. Mitchell wrote the book, but then let it languish. According to Mitchell's obituary in the *New York Times:* "In her Atlanta apartment the manuscript piled up for nine years. Some of it was type written, some of it was scribbled on the backs of laundry lists. It was in desks, bureau drawers and on closet shelves. Friends had read parts of it, but she had never shown it to a publisher." Harold Latham of Macmillan Company read the manuscript in 1935 and had a hunch it could be a best seller. After Macmillan bought the book, it took Mitchell 6 months to weave all the parts of the novel together into a cohesive whole.[7]

> *They say that time changes things,*
> *but you actually have to change them yourself.*
> —Andy Warhol

How long did it take Thomas Jefferson to write the Declaration of Independence?

Just over 2 weeks. Jefferson started work on the seminal document on June 11, 1776, and completed his final draft on June 27. On July 2, the Continental Congress voted for independence, just as the British fleet and army arrived in New York. Congress adopted the Declaration of Independence on the morning of July 4. The first sentence of the final document reads:

> When in the Course of human events, it becomes necessary for one people to dissolve the political bands which have connected them with another, and to assume among the powers of the earth, the separate and equal station to which the Laws of Nature and of Nature's God entitle them, a decent respect to the opinions of mankind requires that they should declare the causes which impel them to the separation.

How long did it take Henry David Thoreau to write *Walden*?

Thoreau spent 2 years, 2 months, and 2 days at Walden Pond, from July 4, 1845, to September 6, 1947. Thoreau's account of his time at the pond, *Walden,* took him nearly 10 years (and 7 or 8 drafts) to complete. *Walden* was finally published in 1854.

> *It is a great art to saunter.*
> —Henry David Thoreau

Time Trivia: How long does a copyright last?

If you write a novel, poem, or a piece of music, the copyright, or ownership, will last your entire lifetime, plus 70 more years. If you wrote a novel in 1970 and died in 2000, the copyright would last until 2070.

If you created the work anonymously, or under a pseudonym, or if the work was made for hire, the copyright lasts for 95 years from the year of its first publication or 120 years from the year of its creation, whichever expires first, according to the U.S. Copyright Office. For works first published prior to 1978, the copyright terms are a bit different; check the Copyright Office website for details.[8]

Music

How long does it take to compose a symphony?

A week, a year, or much, much longer.

Wolfgang Amadeus Mozart, for instance, was a quick worker. He wrote his last three symphonies (39, 40, 41) in a few weeks during the summer of 1788. The speed of his creativity did not affect the quality of his music: Symphonies 39, 40, and 41 are considered some of the finest works of classical music ever made. Over his short, 35-year life Mozart created 600 pieces of music.

Ludwig van Beethoven, however, spent years on his symphonies. He started work on his Ninth Symphony—one of his masterpieces—in 1818 and finished it about 7 years later, in early 1824. Many of his other symphonies were also written over a span of years, rather than weeks.

How long do rock banks last?

Days, months, or until the end of the members' lives. Guess how long the following bands hung together?

 Pop Quiz! Match the band to their lifespan

Beach Boys	30 years (1965 to 1995)
The Beatles	51+ years (1962 to present)
Blind Faith	52+ years (1961 to present, sort of)
Bob Marley and the Wailers	19 years (1964 to 1983; then intermittent reunions)
The Doors	37+ years (1976 to present)
Grateful Dead	8 years (1965 to 1973)
Led Zeppelin	8 years (1965 to 1973)
Rolling Stones	10 years (1960 to 1970)
U2	12 years (1968 to 1980)
The Velvet Underground	7 years (1974 to 1981)
The Who	6 months (1969)

Answers:

> **Beach Boys:** 52+ years (1961 to present, sort of); **The Beatles:** 10 years (1960 to 1970); **Blind Faith:** 6 months (May to October 1969); **Bob Marley and the Wailers:** 7 years (1974 to 1981); **The Doors:** 8 years (1965 to 1973); **Grateful Dead:** 30 years (1965 to 1995); **Led Zeppelin:** 12 years (1968 to 1980); **Rolling Stones:** 51+ years (1962 to present); **U2:** 37+ years (1976 to present); **The Velvet Underground:** 8 years (1965 to 1973); **The Who:** 19 years (1964 to 1983; then intermittent reunions)

4'33" : A composition of time

In 1952, experimental composer John Cage created a three-movement score for "any instrument" that instructed the performer *not* to play the instruments. The score, which lasted for 4 minutes and 33 seconds, specified that the first movement should be 30 seconds; the second, 2 minutes and 23 seconds; and the third, 1 minute and 40 seconds.

Those who attended the premiere of *4'33"* were perplexed, even irked. Cage later explained, "They missed the point. There's no such thing as silence. What they thought was silence, because they didn't know how to listen, was full of accidental sounds. You could hear the wind stirring outside during the first movement. During the second, raindrops began patterning the roof, and during the third the people themselves made all kinds of interesting sounds as they talked or walked out."[9]

> *What happens if we stop, and take the time*
> *to look more carefully? Then the world unfolds like a flower,*
> *full of colors and shapes that we had never suspected.*
> —James Elkins, art critic and historian

How much time do Americans spend playing musical instruments?

Unclear. But according to a 2006 poll, more than half of American households have one person (age 5 or older) who plays an instrument; 40 percent of households have two or more musicians.[10] Why don't *more* kids play instruments? A 2009 Gallup poll found that nearly half of young people said that they don't have the time; 19 percent said sports got in the way; 14 percent said that video games were to blame; and another 14 percent fingered time-consuming extracurricular activities.[11]

While many Americans play music when they're young, they taper off as adults. A survey by the National Endowment for the Arts found that just 13 percent of adults play an instrument.[12]

Film

How long do feature films last?

About 110 minutes.[13] Comedies and children's films tend to be shorter, around 90 minutes, while summer blockbusters drift up to 2 hours. But here's a shocker: over the past century, the top-rated films have actually gotten longer. (Isn't everything getting shorter, faster?) IMDB averaged the running times of the top 50 films in each decade over the last 100 years; here's what they found:

Decade	Average length of the top 50 films
1910s	79 minutes
1920s	98 minutes
1930s	96 minutes
1940s	109 minutes

1950s	114 minutes
1960s	127 minutes
1970s	125 minutes
1980s	129 minutes
1990s	127 minutes
2000s	129 minutes

How long is the longest film?

A few avant-garde films last for days. Buracz Bosnitz, a Czech director, created and screened *Film Stock* (1968), a completely blank film with a running time of 908 hours—a world record. After it was screened, the stock was reused to film other movies.[14] *Modern Times Forever* (2011), a Danish film, runs for 240 hours, or 10 days, which may make it the second-longest film. Some feature films last up to 15 hours, but those are typically released in parts.

How long do film shots last?

Some takes last for seconds; other takes last the entire length of a film. The historical drama *Russian Ark* consists of one 96-minute sequence shot. Action and horror movies tend to have very short average shot lengths, while more ponderous stories, like documentaries, typically have longer average shot lengths. ASLs of films reveal differences in the styles of directors and in the demands of certain genres. Consider the shot lengths of 12 popular films from the past century:[15]

Jack and the Beanstalk, 1902, ASL: 53.2 seconds

Birth of a Nation, 1915, ASL: 7 seconds (187-minute version)

All Quiet on the Western Front, 1930, ASL: 9.4 sec

Psycho, 1960, ASL: 6.2 sec

The Graduate, 1967, ASL: 5.7 sec

Jaws, 1975, ASL: 6.6 sec

Annie Hall, 1977, ASL: 22.3 sec

Scarface, 1983, ASL: 6 sec

Roger & Me, 1989, ASL: 8.2 (Documentary)

The Bourne Supremacy, 2004, ASL: 2.4 sec

The Hangover, 2009, ASL: 4.1 sec

The Artist, 2011, ASL: 5.4 sec

 Time Trivia: Slow-motion timing

Slow motion is the result of fast frames. Directors such as Wes Anderson, Sam Peckinpah, and Martin Scorsese have used slow motion to create exciting film moments. Consider the slow-motion fight scenes from *Raging Bull,* the final shootout in *The Wild Bunch,* or the credit sequence in *Reservoir Dogs*—they are unforgettable, thanks to their artificial elongation. Yet to create those scenes, the camera was actually sped up. Instead of shooting at the normal rate of 24 frames per second (fps), slow-motion scenes are shot at 48 fps, 72 fps, or more; the faster the camera shoots, the slower the scene plays. The opposite is also true: when a scene is shot at 16 fps or 9 fps, the result is sped up, jerky movements such as the scenes in the Beatles movie *A Hard Day's Night.*

The longest-running midnight movie award goes to . . .

The Rocky Horror Picture Show! The R-rated British film, directed by Jim Sharman and starring Tim Curry, Susan Sarandon, and Barry Bostwick, premiered in 1975; its first midnight showing was in 1976.[16] Theaters across the country continue midnight screenings of what some consider the first interactive movie (thanks to audience participation), but one theater stands out, the Oriental Theatre in Milwaukee. Since January 1978 the Oriental has shown the *RHPS* more than 1,700 times.[17]

Time Trivia: Making Harry Potter

It took 14 weeks to create a one-third-size model of the Weasley home, the Burrow, for the film *Harry Potter and the Half-Blood Prince.* After Bellatrix Lestrange and other Death Eaters set the house on fire, it took just 6 minutes for the model house to burn down. Sniff, sniff.

Pop Quiz : How Do Movie Characters Tell Time?

Who says wristwatches are passé? Match the famous character to his or her famous watch. (Answers on the following page.)

Character/Movie	Wristwatch Options
Dumbledore in *Harry Potter*	Citizen PMU56-2373
Tony Montana in *Scarface*	Hamilton Ventura
Mikael Blomkvist in *The Girl with the Dragon Tattoo*	Watch with 12 hands and no numbers
Egon Spengler in *Ghostbusters*	TAG Heuer Carrera Automatic
Dr. Susan McCallister in *Deep Blue Sea*	Seiko Voice Note
James Bond in *Dr. No*	Omega Aqua Terra Mid Size Chronometer
Matt Hooper in *Jaws*	Rolex Submariner 6538
Agent J in *Men in Black*	Citizen PMU56-2373
Cobb in *Inception*	Alsta Nautoscaph Vintage Diver
Erin Brokovich	Timex Ironman Triathlon
Lieutenant Pete "Maverick" Mitchell in *Top Gun*	Swatch Marquita
Harold Crick in *Stranger Than Fiction*	Porsche Design Cronograph

Answers:

Dumbledore in *Harry Potter* wears a watch with 12 hands and no numbers; Tony Montana in *Scarface*, Omega La Magique; Mikael Blomkvist in *The Girl with the Dragon Tattoo*, Omega Aqua Terra Mid Size Chronometer; Egon Spengler in *Ghostbusters*, Seiko Voice Note; Dr. Susan McCallister in *Deep Blue Sea*, Citizen PMU56-2373; James Bond in *Dr. No*, Rolex Submariner 6538; Matt Hooper in *Jaws*, Alsta Nautoscaph Vintage Diver; Agent J in *Men in Black*, Hamilton Ventura Chrono; Cobb in *Inception*, TAG Heuer Carrera Automatics; Erin Brokovich, Swatch Marquita; Lieutenant Pete "Maverick" Mitchell in *Top Gun*, Porsche Design Chronograph; Harold Crick in *Stranger Than Fiction*, Timex Ironman Triathlon

Architecture

How many years did it take to build the Taj Mahal?

The majestic tomb took over 20 years to complete. In 1632, the ruler of the Mogul Empire, Shah Jahan, ordered a tomb to be built for his beloved wife, Mumtaz Mahal, in the city of Agra. A nine-mile-long ramp was built through Agra to transport marble and materials to the construction site. The massive project required 20,000 workers, 1,000 elephants, and numerous oxen to move the heavy materials. The project was finished in 1653.

How quickly was the Empire State Building built?

The building was completed ahead of schedule, taking just 1 year, 45 days, and 7,000,000 man-hours to construct. Work began on March 17, 1930, and was completed on May 1, 1931. Why so speedy? First, a race was on. The builders were determined to create the tallest building in the world; the folks working on the Chrysler Building had the same idea. Second, the sooner the building was completed, the sooner it would bring in revenue. Ironically, the building was mostly unoccupied for nearly a decade, causing locals to name it the Empty State Building.[18]

How long did it take to build the World Trade Center?

5 years

How long did it take for the Twin Towers to fall on September 11, 2001?

The North Tower, 1 World Trade Center, collapsed 1 hour and 45 minutes after being struck by American Airlines Flight 11. The South Tower, which was hit second by United Airlines Flight 175, fell in 56 minutes. The debris from the fallen buildings smoldered for 3 months.

How long does it take to build a bridge?

These days, not long at all. But in the 1800s, when bridge building was new, it took years and often many lives to complete a bridge. Here are some notable structures, ranked by construction time.

Old ideas can sometimes use new buildings.
New ideas must use old buildings.
—Jane Jacobs, *The Death and Life of Great American Cities*

Brooklyn Bridge: 13 years, 5 months
(January 2, 1870, to May 24, 1883)

During the ambitious and lengthy construction project, an estimated 20 to 30 men died, including the original engineer on the project, John Roebling. Roebling contracted a lethal case of tetanus after his foot was crushed by a ferry near the worksite.

Verrazano-Narrows Bridge: 5 years, 3 months
(August 13, 1959, to November 21, 1964)

The bridge was the last public works project overseen by Robert Moses, the New York State parks commissioner. The bridge was controversial (as were quite a few Moses projects)— many families in Bay Ridge, Brooklyn, were moved to make way for the new structure. From 1964 to 1981, the Verrazano-Narrows Bridge held the title as the world's longest suspension bridge (it is still the longest suspension bridge in the United States).

Golden Gate Bridge: 4 years, 5 months
(January 5, 1933, to May 28, 1937)

The gorgeous bridge, which links San Francisco to Marin Country, gained its distinctive color thanks to consulting architect Irving Morrow who was struck by the vibrancy of the reddish-orange primer painted on the steel. Morrow had to convince the bridge's board of directors that orange, rather than gray, was the way to go. He wrote: "The Golden Gate Bridge is one of the greatest monuments of all time. Its unprecedented size and scale, along with its grace of form and independence of conception, all call for unique and unconventional treatment from every point of view. What has been thus played up in form should not be let down in color."[19]

Chesapeake Bay Bridge-Tunnel: 3 years, 6 months
(September 1960 to April 15, 1964)
The 17.6-mile bridge-tunnel complex was completed in just 3.5 years. The creation was named one of the Seven Engineering Wonders of the Modern World in 1964.

Royal Gorge Bridge: 6 months
(June through November 1929)
Suspended 955 feet over the Arkansas River, the bridge was finished in just six months without any major accidents or deaths—perhaps because there was a financial incentive: the Royal Gorge was built as a tourist attraction rather than as a main thoroughfare by the Royal Gorge Bridge & Amusement Company.

Time Trivia: Itapúa Dam

It took 17 years (and $18 billion) to build the Itapúa Dam, one of the Seven Wonders of the Modern World, located on the border of Brazil and Paraguay. Construction started in 1975 and ended in 1991, though the first unit began operating in 1984.

How many seconds does it take to reach the top of a skyscraper?

High-speed elevators can transport visitors 100 stories into the sky in less than a minute. Here's a sampling of buildings with ear-poppingly fast elevators, in order of speed.

Taipei 101 (Taiwan): 39 seconds to the 101st floor
Elevator speed: 1,010 meters per minute
Building height: 509.2 meters (1,671 feet)
Floors: 101

Yokohama Landmark Tower (Japan): 73 seconds to the top
Elevator speed: 750 meters per minute
Building height: 296 meters (972 feet)
Floors: 73

Burj Khalifa—The Tallest Building in the World (Dubai, United Arab Emirates): Less than 60 seconds to the 124th-floor observation deck
Elevator speed: 600 meters per minute
Building height: 828 meters (2,717 feet)
Floors: 160+

John Hancock Building (Chicago): 39 seconds to the 94th-floor observatory
Elevator speed: 549 meters per minute
Building height: 344 meters (1,127 feet)
Floors: 100

Stratosphere Tower (Las Vegas): 30 seconds
Elevator speed: 549 meters per minute
Building height: 350 meters (1,149 feet)
Floors: 113

8 Jin Mao Tower Place (China): 45 seconds to the top
Elevator speed: 540 meters per minute
Building height: 421 meters (1,380 feet)
Floors: 88

Empire State Building (New York): 45 seconds to the 80th floor
Elevator speed: 426 meters per minute
Building height: 443 meters (1,453 feet)
Floors: 103

Time Trivia: Elevator Waits

- Elevator doors are programmed to close more quickly in fast-paced cities like New York and Chicago than in slower-paced cities and Europe.

- The average wait time for an elevator in a typical 16-floor building with a dispatch system—where you type in your floor on a keypad in the lobby and are then sent to a specific elevator—is just 13 seconds.

- The average wait time for an elevator with a conventional system—where you press the up button on a lobby keypad and wait for a car to appear—is 138 seconds.[20]

How long does it take to become an expert?

About 10,000 hours (and yes, that's an oversimplification). Malcolm Gladwell calls this the 10,000-Hour Rule, but this "rule" is based on research by K. Anders Ericsson, a professor of psychology at Florida State University and one of the top researchers in the field of expertise.

In the early 1990s, Ericsson and his colleagues studied the habits of top violinists at the Berlin Academy of Music and found that the musicians who had reached the highest levels of expertise and gained the most international acclaim had spent the most time practicing, about 10,000 hours to be exact.

But it's not just any 10,000 hours of practice that gets you to the top—the practice has to be deliberate. Deliberate practice, as Ericsson calls it, is typically solitary and involves setting goals and concentrating on specific techniques. Usually this process requires a teacher or coach who can design drills or sessions and give useful feedback. "If you put in a lot of hours of not stretching yourself, you won't get much benefit," Dr. Ericsson explained to me in an interview.

The key to success is not about talent, Ericsson believes, it's about practice—lots of it. "There is no evidence that there is innate talent, apart from the effects of body size in sports," he said. "There is surprisingly little hard evidence that anyone could attain any kind of exceptional performance without spending a lot of time perfecting it."

Improvement can come at any age. "In virtually any area, you can take a scientific approach to changing," Ericsson said. Meaning, whatever you love to do, you can improve by practice, practice, practice.

> *All of a sudden the progress will stop one day,*
> *and you will find yourself, as it were, stranded. Persevere.*
> *All progress proceeds by such rise and fall.*
> —Swami Vivekananda, *The Yoga Sutras of Patanjali:*
> *The Essential Yoga Texts for Spiritual Enlightenment*

Energy
Sports & Fitness

Since so many Americans define themselves by their sports, they feel righteous about allocating copious amounts of time to them. Think of the hours that fans devote to watching games, talking about games, and debating the merits of players in the games. Fantasizing about sports has become an organized pursuit. Then there's the time we devote to *our own* sporting routines. We walk, we jog, we go to the gym, we take classes, or we *talk* wistfully about doing all the aforementioned. How much time do we really devote to sports and fitness? How much time should we devote? And what about professional athletic careers: How long do they really last? Read up before your next call to sports-talk radio.

Sports Careers

At what age do athletes reach their peak?

Pro athletes hit the peak of athletic perfection in their mid-20s. Muscular strength overall tends to crest at 25 and then decrease by about 4 percent per decade until the age of 50. Female athletes typically reach their peak a year or so before male athletes.

Researchers estimate the average age of "athletic peakness" (to coin a phrase) by studying the ages at which top athletes win their highest honors. That's how the numbers below were arrived at, but the data are not definitive. Every few years a new study comes out that skews the numbers slightly up or slightly down. What's more, star athletes often beat the averages, staying at the top of their game well into their 30s. There's lots of debate on this topic, particularly when it comes to NFL players.[1]

Sport	Peak age for men	Peak age for women
Baseball (hitters and pitchers)	29	NA
Basketball	24 to 25	NA
Football	24 to 27	NA
Golfing	31	31
Gymnastics	23	17.5
Hockey	25 to 27	NA

Jumping	24	23
Running— short distance	23	22
Running— long distance	27	27
Soccer, field players	28	26
Swimming	19	17
Tennis	24 to 25	23.3

How long do pro careers last?

About 5 to 7 years on average. Measuring career length is a tricky science, and the averages can change depending on the yardsticks used. Also, in some team sports, longevity depends on the specific position of the player; soccer goalies, for instance, have longer careers than field players.[2]

Pro Player	Average career length
Baseball	5.6 years
Basketball	5.5 to 6 years
Hockey	5.5 years
Football	3.5 to 6 years (hotly debated!)
Tennis	7 years

World Record: The longest winning streak

Pakistani squash player Jahangir Khan, who won 555 consecutive games between 1981 and 1986,[3] is considered to be the greatest squash players of all time. His winning streak is one of the longest winning runs by any athlete in professional sports.

Sports Events

How long do games last?

Baseball	MLB games are, of course, timeless. But, the average game lasts 2 hours and 50 minutes, depending on the year and the team.
Basketball	2 hours and 30 minutes
Boxing	A professional match lasts on average 47 minutes (12 rounds x 3 minutes per round + 1 minute in between rounds)
Football	The average NFL games lasts 3 hours and 5 minutes
Soccer	A World Cup soccer match has two 45-minute matches, plus a 15-minute halftime break. A game lasts about 105 minutes, plus a few more minutes for penalties, etc.
Test Cricket	A match lasts up to 5 days
Water Polo	60 to 70 minutes[4]

How long do spectators spend at sporting events?

Between 2.8 and 3.2 hours. By comparison, moviegoers spend 2.2 hours watching a film.[5]

 World Record: Couch Potato Time

Jeff Miller, 26, from Rogers Park, Illinois, watched televised sports for 72 hours nonstop in January 2010. Miller set the world record for the longest time spent watching sports on TV. Go, Jeff!

Football: lots of airtime, not much game time

An NFL game eats up 174 minutes of broadcast time, yet the average amount of ball time is a mere 11 minutes, according to a *Wall Street Journal* study of four televised games. The *WSJ* study found that the football is in play on the field for an average of 11 minutes per game and that a typical play lasts just 4 seconds. Additional highlights from the *WSJ* study:[6]

- Standing around time: 67 minutes

- Replay time: 25 minutes

- Commercial time: nearly 60 minutes

Baseball: lots of airtime, slightly more game time (than football)

Surprisingly, sports fans watching a MLB game on TV see 3 *more* minutes of action than those watching an NFL game. A nine-inning game lasts about 128 minutes and has 14 minutes of action, according to a *WSJ* study.

- Standing around time: 88 minutes

- Replay time: 10 minutes

- Commercial time: 42 minutes and 10 seconds

We didn't lose the game; we just ran out of time.
—Vince Lombardi, U.S. football coach

Time Trivia: NFL prep time

The typical NFL season requires 514,000 hours of labor per team—a number that has nearly doubled in the last 20 years, according to a study by the Boston Consulting Group.

More sleep = Better performance

Everyone knows sleep is important for the brain and the body. But sleep is especially important for athletes who want to reach peak performance. One study found, for instance, that when college basketball players slept 10 hours a day, instead of their usual 6 to 9 hours, they were able to run faster and shoot more accurately. The players in the study also reported that they felt less fatigued overall and that their practices and games improved. The authors of the study, which was conducted at the Stanford Sleep Disorders Clinic and Research Laboratory and published in the journal *Sleep,* wrote: "Athletes may be able to optimize training and competition outcomes by identifying strategies to maximize the benefits of sleep." Translation: more sleep = better performance.[7]

Sporting Endurance

For how long can a surfer ride a wave?

Surfers can ride small waves for 5 to 30 seconds, big waves for a minute or more.[8] Gary Saavedra holds the record for the longest time spent surfing on the open water: 3 hours and 55 minutes on the Panama Canal. But he sort of cheated. The Panamanian native surfed behind a MasterCraft boat, which created a constant wave for him to ride.[9]

How long can a swimmer paddle in frigid water?

It gives me the chills just contemplating this act of bravura (insanity?). British lawyer Lewis Gordon Pugh set a record when he swam for 18 minutes and 50 seconds in North Pole water that ranged in temperature from 29°F to 32°F. Pugh had completed long distance swims in every ocean on the earth, but said his 1 km North Pole swim was his most challenging. "I was in excruciating pain from beginning to end and I nearly quit on a few occasions," he said. Why do it? Pugh plunged into the icy water to bring attention to global warming.[10]

How long did Philippe Petit walk the tightrope between the Twin Towers?

Forty-five minutes. On Wednesday, August 7, 1974, in the early morning hours as commuters were making their way to work, Philippe Petit stepped off the South Tower and onto a steel cable he had secretly rigged with the help of a few accomplices. He made eight artful crossings between the two towers. Petit, a French-born high-wire artist—then 25 years old—sat, lay down, and turned corners a quarter mile in the air. If you missed Petit's high-wire act, you can see it in the documentary *Man on Wire*.

How long does it take to run to the top of the Empire State Building?

Nine minutes and 33 seconds, but only if you are Paul Crake. In 2003, the 26-year-old Australian ran up the 1,576 stairs of the Empire State Building in record-breaking time. No other runner has broken the 10-minute mark during the Empire State Building Run-Up, which began in 1978.[11]

How long do dance marathons last?

The world record is 214 days. Dance marathons started in the 1920s and became popular during the Depression as a way for folks to both earn money and fill up time. Contestants danced around the clock but were typically given 15-minute cot rests each hour and 2-minute bathroom breaks every 2 hours. Spectators paid about 25 cents to watch the events and could stay as long as they liked. The record was achieved by Mike Ritof and Edith Boudreaux who danced the days and nights away for 5,152 hours and 48 minutes from August 29, 1930, to April 1, 1931, at the Merry Garden Ballroom in Chicago. The couple's prize money of $2,000 for 214 days' work came out to about 39 cents an hour—in today's dollars that's about $30,000.

How long does it take for a pro golfer to get dressed?

If you're paid by Nike to look sharp, it takes about 6 months. "We work with all of our athletes six months in advance of a given tournament, sharing our direction and thoughts on the right types of style, material, innovation and look [sic] are best suited for the playing condition the athlete is most likely to encounter," a post on Nike's blog explained in 2011. "We then provide a detailed 'scripting sheet' that details each outfit for the four days of competition. This includes not only the 'game day' uniform of polo and pant, but goes down to the detailed level of headwear, cover up, shoe and even the belt as an integral part of that uniform."[12]

Fitness—Adults

How much do Americans exercise each day?

Here's one way to look at it: Americans exercise, on average, for about 20 minutes a day—that average includes everyone, from fanatical gym rats to couch-loving sloths.

Here's another: Just 21 percent of American adults meet the government's goals for weekly aerobic and muscle-strengthening times (see below for specifics). Men exercise more than women. Younger people exercise more than older folks. Whites exercise more than blacks and Hispanics.[13]

And yet another: A recent study, which used government survey data, figured out how much time people in very good health spent exercising based on age, sex, and weight.[14] Check out the table below:

Americans in very good health	Daily exercise time
Overall average	25.9 minutes
Men	27.4
Women	15.1
15- to 24-year-olds	38.7
25 to 54	15.7
55+	20.3
Normal weight	20.9
Overweight	18.1
Obese	15.8

How much time *should* adults spend exercising?

Ideally, those over 18 years of age should work out for 150 minutes each week and do muscle-strengthening exercises 2 or more days a week, according to guidelines published by the Centers for Disease Control and Prevention. But the CDC offers three options:[15]

1. Exercise at a moderate pace for 2 hours and 30 minutes every week *and* do muscle-strengthening activities on 2 or more days a week that work all major muscle groups (legs, hips, back, abdomen, chest, shoulders, and arms).

2. Exercise vigorously (running, jogging, power yoga) for 1 hour and 15 minutes (75 minutes) every week *and* do muscle-strengthening activities on 2 or more days a week that work all major muscle groups.

3. Mix it up. Do some moderate- and vigorous-intensity aerobic activity that equals the guidelines above *and* muscle-strengthening activities on 2 or more days a week that work all major muscle groups.

Good news: You can break your exercise into 10-minute chunks of moderate to vigorous activity. If you go for a brisk, 10-minute walk 2 times a day, every day (140 minutes in all) you will just about have met your weekly fitness quota. One minute of vigorous-intensity activity is about the same as 2 minutes of moderate-intensity activity.

When is the best time to exercise?

In general, the optimal time to work out is late in the afternoon between 4 and 7 P.M.; that's when the body is in its best shape for physical activity:

- Body temperature is at its peak between 6 P.M. and 8 P.M., which means your muscles are warm and flexible and your pain threshold is high. Reaction time and speed improve when your body is warm.

- Positive moods are highest in the evening. And it's usually easier to get motivated to exercise when your spirits are high.

- Muscle strength peaks between 5 P.M. and 9 P.M. One study found that athletes who trained in the evening gained 20 percent more muscle strength than those who trained in the morning.

- Hand-eye coordination is sharpest in the early evening, which is especially important if you're playing racket sports, basketball, or baseball.

- Endurance is best. Aerobic capacity is about 4 percent higher in the afternoon.

How long does it take to change your body?

If a person is a bit doughy and out of shape, it takes from six months to a year to start to see a real change in the body. During that time, a person should be on a rigorous self-improvement program: working out, lifting weights most days of the week, and eating well every day.

Time Exercise 6: Count Your Steps

If walking is your favored form of exercise, try this: count the number of steps you take in a minute. Or better yet, get a pedometer, clip it to your belt, and then walk for 1 minute and check out the total number of steps taken. Your goal should be 100 steps per minute. That's the pace that is considered a moderate-intensity workout, according to a study that appeared in the *American Journal of Preventive Medicine* in 2009.

If you want to meet exercise guidelines set by the CDC, aim to walk a minimum of 3,000 steps in 30 minutes, 5 days a week. The study's authors note: "Three bouts of 1,000 steps in 10 minutes each day can also be used to meet the recommended goal."[16]

Fitness–Kids

How much time do children spend exercising?

Not as much as they should. High school kids are supersedentary. Just 11 percent of teen girls and 25 percent of boys get the suggested 60 minutes of exercise a day (see below). Younger kids are more active. Most 9- to 13-year-olds, about 77 percent, get regular daily exercise.[17]

How much time *should* children spend exercising?

The government recommends 60 minutes a day of moderate-
or vigorous-intensity aerobic activity for kids; part of that hour
should include muscle-strengthening and bone-strengthening activities at least 3 days a week.

 Time Trivia: Sporting time is good for the brain

Children who spend more time playing sports or participat-
ing in structured activities tend to have higher math scores
and fewer behavior problems than their less sports-oriented
classmates.

Exercise and Health

More hours sitting = less hours of life

"By too much sitting still, the body becomes unhealthy and soon
the mind," wrote Henry Wadsworth Longfellow in the nineteenth century. Longfellow was a prescient guy.

Americans spend a lot of time on their derrieres. One study
found that adults spend 8.5 hours, on average, sitting each day.
"No worries," you may think. "I go to the gym every morning
and work out for an hour." Alas, even daily workouts won't
counteract the deleterious effects of sitting. Research by the

American Cancer Society found that women who spent more than 6 hours a day sitting were 37 percent more likely to die during the study period than those who sat fewer than 3 hours a day. Men who sat more than 6 hours a day were 18 percent more likely to die than those who sat fewer than 3 hours per day. The connection between sitting and mortality stayed constant even when the researchers adjusted for physical activity level. "Prolonged time spent sitting, independent of physical activity, has been shown to have important metabolic consequences," wrote one of the study's authors, Alpa Patel, Ph.D, "and may influence things like triglycerides, high density lipoprotein, cholesterol, fasting plasma glucose, resting blood pressure, and leptin, which are biomarkers of obesity and cardiovascular and other chronic diseases."[18]

If you sit at a desk all day, get up every 20 to 30 minutes to stretch and walk, even if it's just to circle around your desk a few times. When you talk on the phone, stand up and pace. Have walking meetings, rather than traditional sitting around a conference table ones. Bottom line (pun intended): don't take life sitting down.

More exercise = longer life

Sitting, bad. Moving, good. The more time you spend being physically active, the longer you will live (up to a point). A landmark research project followed nearly 17,000 Harvard alumni for 16 years and found that death rates were one-fourth to one-third lower among alumni who expended 2,000 or more kcal during exercise each week. Regular exercise reduced coronary death rates by 25 percent to 33 percent. Even when the researchers took into consideration hypertension, cigarette smoking, extremes or gains in body weight, or early parental death, alumni mortality rates were significantly lower among the physically active.[19]

 Time Exercise 7: More Exercise, More Years

Each hour of vigorous exercise that you do will add 2 hours to your life. If you exercise 4 days a week for 1 hour for the next 30 years, you will increase your life expectancy by about 1.4 years. To find out how many years or months your exercise routine will add to your life, Google: Increased Life Expectancy Calculator + BizCalcs.

Make use of time, let not advantage slip.
—William Shakespeare

How many calories can you burn in an hour?

That depends on your weight and the vigor of the activity. Consider these general guidelines for a 125- and 150-pound person.[20]

Activity	125 pounds	150 pounds
Basketball		
• **Full court**	622 calories	747
• **Shooting baskets**	255	306
Biking		
• **12–14 mph**	495	594
Frisbee	172	207
Golf		
• **Carrying clubs**	345	414
• **Using a cart**	195	234
Jumping rope	570	684
Pilates		
• **Beginner**	210	252
• **Intermediate**	292	351
• **Advanced**	360	432

Running		
• 6 mph	570	684
• 8 mph	765	918
• 10 mph	1,020	1,224
Skiing		
• Downhill	495	594
• Cross-country	645	774
Swimming		
• Moderate	345	414
Walking		
• 2 mph	158	198
• 3 mph	248	297
• 4 mph	292	351
• upstairs	458	549
Weight lifting	195	234
Yoga		
• Hatha	158	189
• Ashtanga/Power	292	351
• Bikram/Hot yoga	398	477

Destruction
Wars, Law Enforcement & Crime

Humans don't always behave well. They start wars, they steal, they kill, and they subject others to their cruelty; they act aggressively, deviously, and selfishly. As long as humans populate the earth, there will likely be crimes and punishments. Sometimes destruction is necessary (though I wish I could believe otherwise). The United States might not exist if early settlers had not been ready to wage a war; we might all be living in a different world if D-Day hadn't unfolded as planned.

What does all this have to do with time? The mayhem and destruction that changes the course of history can be startlingly variable in its speed, from the six seconds in Dallas that killed JFK and the flash of the fireball in Hiroshima, at one extreme, to the Hundred Years War on the other. And the fallout from humankind's mayhem can take many lifetimes to correct.

Time at War

How much time has the United States spent at war?

Two-hundred-plus years. The U.S. military has been engaged in some type of offensive for nearly every year since the country was founded—the majority of these wars, though, were on foreign soil. The United States has fought with Native Americans, Brits, Spaniards, numerous pirates, Samoans, Koreans, Germans, Japanese, Iraqis, and on and on.

How long do American wars last?

Here's a rundown of the major American wars, the time they took, and the approximate military lives they claimed. The American Civil War still ranks as the most lethal. Scholars recently revised the death toll upward, from about 620,000 to 750,000—possibly as high as 850,000.

American Revolutionary War: 8 years (1775 to 1783)
  Military lives lost: 25,000 American soldiers

War of 1812: 3 years (1812 to 1815)
  MLL: 20,000

Mexican-American War: 2 years (1846 to 1848)
  MLL: 13,283

American Civil War: 4 years (1861 to 1865)
  MLL: 750,000 to 850,000

Wounded Knee: (December 29, 1890)
   MLL: 153+ Lakota, 25 U.S. soldiers

World War I: 4 years (1914 to 1918)
   MLL: 116,500

World War II: 6 years (1939 to 1945)
   MLL: 405,399

Cold War: 46 years (1945 to 1990)
   MLL: NA

Vietnam War: 20 years (1955 to 1975)
   MLL: 58,200

The War in Afghanistan: 2001 to present
   MLL: 2,000+

Iraq War: 2003 to 2011
   MLL: 4,430

The Shortest and Longest Wars

The shortest war on record lasted 45 minutes. This remarkably quick skirmish occurred between the United Kingdom and Zanzibar on August 27, 1896. A coup d'état in Zanzibar annoyed the Brits who sent the Royal Navy and Marines to try to oust the self-appointed sultan, Khalid bin Bargash, from power. Shelling lasted about 45 minutes before the sultan retreated to the German consulate and the war abruptly ended. The longest war is believed to be the Hundred Years War between England and France, though technically it was a series of battles spanning 116 years (1337 to 1453).

Time Trivia: So you want to be a Navy SEAL?

U.S. Navy SEALs (sea, air, and land) have perhaps the most demanding military job in the world. Because the assignments are so tough, candidates are put through a rigorous Physical Screening Test to determine whether they have the chops for the work ahead. Part of the test has to do with completing physical feats in record time:

- Swim 1,500 feet in under 12 minutes, 30 seconds

- Do 42 push-ups in under 2 minutes

- Do 50 sit-ups in under 2 minutes

- Run 1.5 miles in boots in under 11 minutes, 30 seconds

If a candidate passes the PST, there are other hurdles to jump, including enduring hell week in which SEALs only sleep 4 hours over 5 days, while being pushed to their physical and emotional limits.

The SEAL motto: "The Only Easy Day Was Yesterday."

There was never a good war, or a bad peace.
—Benjamin Franklin, American statesman

Sinking Times

A startling number of ocean liners have come to bad ends—some were victims of war, some of natural disasters. How long does it take for a liner to go down? Consider the fates of these great vessels.

Ocean Liner (weight in gross tons)	Incident	Lives lost	Sinking time
Andrea Doria (29,100)	The Italian vessel collided with the Stockholm liner at 11:10 P.M. on July 25, 1956, near Nantucket. It's believed that the efficiency of the ship's design helped it to stay afloat for over 11 hours after the collision.	46	11 hours
Titanic (46,239)	Struck an iceberg at 11:40 P.M. on April 14, 1912. It was the ship's maiden voyage and various luminaries were on board including Margaret "Molly" Brown and John Jacob Astor IV.	1,517	2 hours and 40 minutes
RMS *Lancastria* (16,243)	Bombed by German planes during WWII at 3:48 P.M. on June 17, 1940. The British ship sunk near the port of St. Nazaire.	1,738	20 to 60 minutes

Britannic (29,100)	Struck a mine off the Greek island of Kea at 8:12 A.M. on November 21, 1916.	30	55 minutes
Lusitania (30,396)	A German torpedo was fired at the ship at 2:09 P.M. on May 7, 1915	1,153 to 1,195	18 minutes
Empress of Ireland (14,191)	Collided with the Storstad at 2 A.M. on May 29, 1914, on the St. Lawrence River	1,014	14 minutes

Prison Times

How long is the average prison term?

American sentences are some of the toughest in the world. No other country puts so many criminals behind bars and many pundits believe that our system is too harsh. "Our sentencing policies are far out of line in terms of cost-effectiveness or compassion, and often do more harm than good," Marc Mauer, executive director of the non-profit organization, *The Sentencing Project,* explained to me by e-mail. "What's more, research has consistently found that any deterrent effect of the justice system is more a function of the certainty of punishment, rather than the severity,"

State prison terms

There were approximately 1.6 million people in U.S. prisons in 2010, and the vast majority of those prisoners, 1.4 million, were in state prisons.[1] What were they locked up for? In 2009 (last available data), slightly more than half of state inmates were serving time for violent offenses, 20 percent were in for property crimes, 18 percent for drug crimes, and 9 percent for public order offenses. The average sentence length was about 4 years, but the time actually served was just 2.

Federal prison terms

Less than 10 percent of the prison population is in federal lockup. In 2010, half of federal inmates were serving time for drug offenses, 35 percent for public-order offenses (largely weapons and immigration), and less than 10 percent each for violent and property offenses.

Because crimes tried in federal court are typically more severe than those tried in state courts, federal sentences tend to be longer. A Department of Justice spokesperson explained to me by e-mail, "Federal offenses labeled robbery are almost exclusively bank robberies, while state robbery offenses seldom include those of banks. Similarly, large-scale international drug crime characterizes a relatively large fraction of federal drug trafficking cases, but few state cases. Federal weapons offenses may entail importation or manufacture of large quantities of weapons, while state weapons offenses typically involve a single firearm."[2]

Decide for yourself whether the punishment fits the crime.

	Mean sentence length for felons sentenced to prison	
Primary offense	in *state* courts	in *federal* courts
Violent offenses		
Murder	244 months	124 months
Sexual assault (including rape)	106	176
Lesser sexual assaults, such as statutory rape or incest with a minor	138	182
Robbery	87	105
Aggravated assault	41	53
Other violent crimes, such as negligent manslaughter and kidnapping	38	150
Property offenses		
Burglary	41	31
Larceny	22	20
Motor vehicle theft	19	43
Fraud/forgery, including embezzlement	28	29

Drug offenses		
Drug possession	23	48
Drug trafficking	38	87
Weapons offenses	32	88
Other nonviolent crimes such as vandalism and receiving stolen property	24	34
Average prison sentence	38 months	65 months
Mean time actually served	2 years[3]	NA

Time Trivia: Serving Life

Only .3 percent of felons were given life sentences by state courts in 2006. However, nearly 1 out of 4 convicted murderers were sentenced to life in prison.

The degree of civilization in a society can be judged by entering its prisons.
—Fyodor Dostoevsky, Russian writer

How long does it take for felons to be sentenced?

After an arrest, it takes about 265 days for a person convicted of a felony in state courts to be sentenced. Those convicted of murder have the longest waits, about 505 days, while those convicted of larceny have the shortest, a median of 220 days.

For how long are committed juveniles locked up?

An average of 148 days (nearly 5 months). Or, if measuring by the median, half are held in placement for more than 112 days, half for less.

There were nearly 93,000 offenders in residential youth facilities on a given day in February 2006 (the last date for which data are available), down from 105,000 in 1997.

How many years do white-collar criminals get?

Draconian sentences are not just for cannibals and serial murderers. Over the past decade white-collar criminals have been given stiff terms, and some of the worst offenders may live out the rest of their lives in jail. Here's a sampling of some of the longest sentences awarded over the past decade.

Incidentally, the term "white-collar crime" was coined by Professor Edwin Hardin Sutherland in 1939 who defined it as "a crime committed by a person of respectability and high social status in the course of his occupation."

Megasentences for White-Collar Criminals

The criminal	The crime	The sentence (sentencing date)
Sholam Weiss	Weiss was convicted of fraud and money laundering, acts that led to the demise of National Heritage Life Insurance. At the end of his trial, Mr. Weiss fled to Austria and was a fugitive for one year. His punishment included a $300 million fine.	845 years (February 2000)
Norman Schmidt	Schmidt ran a fraudulent high yield investment scheme through which he pilfered money to buy race cars, a race track, and other racing-related properties.	330 years (April 29, 2008)
Bernard Madoff	The Ponzi scheme architect admitted to swindling thousands of investors out of billions of dollars.	150 years (June 29, 2009)
Lawrence Duran	As the co-owner of American Therapeutic Corporation, Duran defrauded Medicare of $205 million. Justice officials said the 50-year sentence "reflects the reprehensibility of the defendant's conduct, and is a powerful warning sign to others inclined to cheat the Medicare program."	50 years (September 16, 2011)

Lee B. Farkas	The former chairman of the mortgage firm Taylor, Bean & Whitaker, led a $3 billion fraud involving fake mortgage assets.	30 years (June 30, 2011)
Sholom Rubashkin	Rubashkin was convicted of 86 counts of financial fraud, including bank fraud, mail and wire fraud, and money laundering.	27 years (June 2010)
Bernard J. Ebbers	The former WorldCom CEO was convicted of false financial reporting, and defrauding investors out of $11 billion. He famously drove himself to prison in his Mercedes.	25 years (July 25, 2006)
Jeffrey Skilling	As the CEO of Enron, he engaged in fraud, insider trading, and other crimes that led to the collapse of the company. In addition to his sentence, Skilling was fined $45 million.	24.3 years (October 2006)
Samuel Israel III	He was sentenced to 20 years in prison for defrauding investors of his Bayou hedge fund, plus an additional 2 years when he tried to fake his own suicide.	22 years (April 2008 and July 2009)

Marc Dreier	Dreier, founder of law firm Dreier, LLP, sold more than 85 phony promissory notes to at least 13 hedge funds defrauding them of more than $400 million; he was also accused of stealing from his clients	20 years (July 13, 2009)
John and Timothy Rigas	The father and son team embezzled $2.3 billion from their cable company, Adelphia Communications. The pair (John is the dad) was convicted of conspiracy.	Timothy: 20 years John: 15 years (June, 2005)
L. Dennis Kozlowski & Mark H. Swartz	Accused of defrauding investors of $400 million, the former Tyco CEO and his lieutenant were convicted of fraud, conspiracy, and grand larceny.	8.3 to 25 years (September 19, 2005)

Are longer sentences more punitive?

Maybe. Maybe not. Humans adapt remarkably well to new situations, whether good or bad. Researchers have found that situations that promise happiness, such as marriage and children, typically create a surge in well-being which fades with time. The same is true for unhappy situations, like imprisonment. "To a noteworthy degree, people adapt to being in prison. Their happiness drops at the beginning and they expect it to remain at that low ebb, but it rebounds impressively as they adjust to their new surroundings," explains law professor John Bronsteen and his colleagues in their paper, "Happiness and Punishment." What's more, the authors write, "Contrary to expectations, adjusting the size of a fine or the length of a prison sentence does not meaningfully adjust the amount of unhappiness that is ultimately experienced by the offender."[4]

Prison itself may not the worst part of being sentenced. Bronsteen et al. continue, "virtually any period of incarceration, no matter how brief, has consequences that negatively affect prisoners' lives in ways that resist adaptation, even after they have been released. Prisoners are often abandoned by their spouses and friends, face difficulty finding and keeping employment, and must grapple with incurable diseases contracted during their incarceration. Thus, living in prison itself becomes less oppressive with time, but the effects of having been in prison tend to linger and to diminish happiness indefinitely."

Indeed, 2 weeks after being released from prison, former inmates are 12.7 times more likely to die than people of the same age, sex, and race in the general population. The leading cause of death among former inmates is a drug overdose.[5]

Trial Times

How long do criminal trials last?

Lengthy trials not only wreak havoc on the innocent folks involved with them, but also cost tax payers millions of dollars. Let's look at some of the longest criminal trials in recent U.S. history.

The McMartin Preschool Trial: 30 months

The trial was one of the longest, costliest, and wackiest in U.S. history. The case dragged on for years, cost taxpayers $13 million, and included wild reports of satanic rituals, horse killings, and sodomy. "So much agony for so little," the *New York*

Times proclaimed after the tortuous case came to a close in January of 1990 and the preschool director, Peggy McMartin Buckey, and her son, Raymond Buckey, were acquitted on 52 counts of molesting children at their school.[6] Mr. Buckey went through a second trial, which lasted 3 months and ended in a mistrial.

The Hillside Strangler Trial: 23 months

The Strangler turned out to be two men, Kenneth Bianchi and Angelo Buono, cousins and fellow serial killers who terrorized and killed women in Los Angeles between 1977 and 1978. Bianchi pled guilty and testified during the trial; both men were sentenced to life without parole on October 31, 1983.

Lucchese Mob Trial: 21 months

Twenty-one members of the Lucchese crime family were accused of selling and distributing cocaine, credit-card fraud, gambling, and loan sharking in New Jersey and Florida. After a nearly 2-year trial and 14 hours of deliberation, a jury found the defendants not guilty on August 26, 1988, and the men were acquitted.

Pizza Connection Trial: 17 months

It started with an international Mafia plot to distribute heroin and cocaine and launder the profits through a network of pizza parlors. It ended, after a lengthy 17-month trial, with 18 men (many of whom did not speak English) convicted of running an international ring that distributed millions of dollars' worth of drugs. The trial began on October 24, 1985, and ended on March 2, 1987, making it the longest criminal jury trial in Manhattan and one of the costliest at $50 million.

The Night Stalker: 14 months

Richard Ramirez, a.k.a. the Night Stalker, was sentenced to death in Los Angeles County on September 20, 1989, after a long and bizarre trial. Ramirez, a devil-worshipping Texas drifter, gained fans among other Satan worshippers, who attended his trial each day. During the trial, one of the jurors was murdered by her boyfriend while fighting with him about the case. In response to his 19 death sentences, the remorseless serial killer quipped: "Big deal. Death always went with the territory. I'll see you in Disneyland."

The OJ Simpson Murder Case: 9 months

The trial seemed to go on forever. While it was by no means the country's longest trial, it was certainly the most publicized. The trial started on January 29 and ended on October 3, 1995. After months of riveting testimony, "The Juice" was found not guilty of killing his ex-wife, Nicole Brown-Simpson, and her friend, Ronald Goldman.

How long did the Michael Jackson trial last?

Not long, when you consider his celebrity. The criminal court session lasted six weeks. In the end, Jackson's doctor, Conrad Murray, M.D., was convicted of involuntary manslaughter in November 2011 and sentenced to four years in prison.

Crime Times

When do robbers rob banks?

The most popular robbing times are Fridays between 9 A.M. and 11 A.M. Of the 5,628 bank robberies that occurred in 2010, 20 percent took place on Friday, and 28 percent took place in the morning. Why so early? According to the Department of Justice, robbers choose the morning because banks are less populated and are also flush with cash from early deliveries and emptying of night safe and deposit boxes. Why Friday? The DOJ surmises that addicts may try to load up with cash on Friday so they will have funds to buy supplies for the weekend.[7]

How long did Bonnie and Clyde's crime spree last?

Just 21 months. Bonnie Parker and Clyde Barrow met in Texas in January 1930. Bonnie was 19 and married at the time to an imprisoned murderer. Clyde, 21, was jailed soon after their first meeting. Their joint life of robbing, killing, thieving, and kidnapping began in 1932, when Clyde was paroled from jail. The FBI started its hunt for the notorious pair in mid-1933 and their life of crime ended on May 23, 1934, when they were ambushed and killed by police officers. The couple was believed to have murdered 13 people.

Dictator Times

2011 was a bad year for despots. The Arab Spring took down a few ruthless leaders, while others succumbed to natural causes or their own reckless behavior.

Dictators typically stay in command for decades, wreaking havoc on their countries and amassing untold wealth for their families. Some leave greatly enriched (who can forget Imelda's shoes); a few don't make it out alive.

Can you guess how long these menaces lasted?

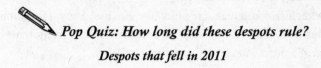

Pop Quiz: How long did these despots rule?

Despots that fell in 2011

Zine Al-Abidine Ben Ali of Tunisia: ____?____ years
The Arab Spring began in December 2010 when one angry fruit vendor in Tunisia lit himself on fire. Within a month of the vendor's protest, Ben Ali was ousted by what *Newsweek* magazine called a "Twitter-fueled youth revolt." He was sentenced in absentia to 35 years in prison for embezzlement and misusing public funds. In 2012, Ben Ali was also convicted of torturing army officers over an alleged coup plot.

Hosni Mubarak of Egypt:____?____ years
Next to fall: the leader of Egypt. After 18 days of popular protests, Mubarak reluctantly stepped down from his decades-long presidential rule in February 2011.

Muammar Gaddafi of Libya:____?____ years
After Mubarak: Gaddafi. Protests against Gaddafi's rule began in early 2011, but the stubborn colonel refused to cede power. He was finally captured and killed by the Libyan National Liberation Army in October 2011.

Ali Abdullah Saleh of Yemen:____?____ years
Saleh didn't give up easily either. Yemenites called for his resignation in early 2011, but the Yemen leader endured a year of nonstop violence and an assassination attempt before finally ceding power to his deputy on February 27, 2012.

Kim Jong-il of North Korea:____?____ years

The reclusive and menacing dictator was best known for human rights abuses and destroying his country's economic system. He died of a heart attack on December 17, 2011.

Laurent Gbagbo of Ivory Coast:____?____ years

Talk about tenacious. After his opponent was elected to office, Gbagbo refused to step aside and even stayed on after economic sanctions were imposed on his country and a civil war erupted. He was finally arrested in April 2011 and then charged by the Hague with crimes against humanity (including 3,000 murders, rapes, and mistreatment of his citizens).

Despots that fell before 2011

Saddam Hussein of Iraq:____?____ years

He was as ruthless as he was ambitious. The egomaniacal ruler used poison gas on his citizens, engaged in expensive, futile wars, and executed detractors, including his own family members. He called himself president until the very end. Hussein was sentenced to death by an Iraqi court and hanged in December 2006.

Saparmurat Niyazov of Turkmenistan:____?____ years

Bossy and eccentric, the totalitarian tyrant named months of the year after family members, banned the use of lip syncing at public concerts, and shut down all Internet cafés. He died in office at age 66.

Slobodan Milošević of Serbia/Yugoslavia:____?____ years
Where to begin? Milošević was indicted by the International Criminal Tribunal for the former Yugoslavia (ICTY) for: "The forced deportation of approximately 800,000 Kosovo Albanian civilians; Sexual assaults carried out by forces of the Federal Republic of Yugoslavia and Serbia against Kosovo Albanians, in particular Women; and The extermination or murder of hundreds of Croat and other non-Serb civilians," among other equally horrific charges. He died in his cell at the United Nations detention center on March 11, 2006, before being sentenced.

Augusto Pinochet of Chile:____?____ years
General Pinochet seized power in a military coup supported by the United States, which overthrew Salvador Allende's Marxist government. During the general's terrorist rule, more than 3,200 people were executed or disappeared, and thousands were detained and tortured or exiled.

Ferdinand Marcos of Philippines:____?____ years
An authoritarian and corrupt leader, Marcos created a cult of personality that helped keep him in power. In 1983, his government was implicated in the assassination of his political opponent, Benigno Aquino Jr., which led to his downfall. He fled the country for Hawaii in 1986. In 1988, he was charged with embezzling more than $100 million from the Philippine government, but he had left some of his wealth behind: "An inventory at Malacanang Palace in Manila found more than a thousand pairs of shoes belonging to Mrs. Marcos, 888 handbags, 71 pairs of sunglasses and 65 parasols—part of a lavish inventory that shocked and offended the world more than any past estimates of the Marcos's enormous wealth," according to the *New York Times*.[8]

Pol Pot of Cambodia:____?____ years

The Khmer Rouge regime headed by Pol Pot combined extremist ideology with ethnic animosity and a "diabolical disregard for human life to produce repression, misery, and murder on a massive scale," according to the Cambodian Genocide Program. During his rule, citizens were executed, tortured, and starved, resulting in the deaths of approximately 21 percent of the Cambodian population. He died in 1998 while under house arrest.

Answers:

Ben Ali, 24 years (1987 to 2011); Mubarak, 30 years (1981 to 2011); Gaddafi, 42 years (1969 to 2011); Saleh, 21 years (1978 to 2011); Kim Jong-il, 17 years (1994 to 2011); Gbagbo, 11 years (2000 to 2011); Hussein, 30+ years (1970s to 2006); Niyazov, 16 years (1990 to 2006); Milošević, 11 years (1989 to 2000); Pinochet, 17 years (1973 to 1990); Marcos, 21 years (1965 to 1986); Pol Pot, 19 years (1975 to 1997).

We are what we repeatedly do.
—Aristotle

Money
Currency & Investing

Time is our most precious resource, but money comes in a close second. Money can, after all, buy better health care, better education, and better housing. The relationship between time and money is interdependent, perhaps even codependent. Time *is* money.

Whoa. This stuff is cosmic. To bring it back down to earth, I consulted one of my favorite thinkers, Benjamin Franklin. Ben had a way of cutting to the chase with words that were simple, sage, and kind. In his autobiography, Franklin wrote about managing time and money in a way that is still relevant in the twenty-first century. He advised: *Lose no time; be always employ'd in something useful; cut off all unnecessary actions.* On money he's more emotional. *Make no expense but to do good to others or yourself; i.e. waste nothing.*

You will see in this chapter how the factor of time influences the way we earn, spend, and invest our money.

CURRENCY

Speaking of Benjamin Franklin, the founding father was one of the chief advocates of paper money back in the mid-1700s. He wrote about paper money's use, designed bills, and printed currency on his own press. What couldn't Franklin do?

Consider this basic information on the bills and coins we handle each day.

How long does currency stay in circulation?

Paper money is lightweight, but it wears out. A one-dollar bill has a life span of less than five years, while a coin lasts for twenty-five. How does the government replace bad bills with crisp ones? When banks come across worn or soiled bills they send them to a local Federal Reserve bank, and the Fed gives them crisp new notes in exchange. Damaged coins must be sent back to the U.S. Mint in Philadelphia.[1]

Bill	Life Span
$ 1	4.8 years
$ 5	3.8
$ 10	3.6
$ 20	6.7

$ 50	9.6
$100	17.9

How many bills are printed in one day?

The Bureau of Engraving and Printing prints 16.6 million dollar bills each day.

How long do coin designs endure?

The Secretary of the Treasury has the authority to change designs after 25 years. Congress, however, can authorize a change earlier.

Making Money Grow

Now on to the fun stuff!

How long does it take to become a millionaire?

That depends on how much you are willing to save. If you deposit $500 into an investment that earns 8 percent annually each and every month, you'll have $1,000,000 in 34 years. Put $500 in a more conservative savings account that earns 2 percent, and you'll reach your goal in 74 years.

What if you want to get rich faster? What if you need $1 million in 15 years? Then you have to sock away $3,000 a month and pray that it earns 8 percent annually.

How long will it take for your 401(k) to make $1 million?

This formula is a lot more satisfying. Say you are 35 years old, earn $45,000 annually, and put 10 percent of your income in your 401(k) each year (about $375 a month). Assuming that your employer match is 50 percent, you receive a 4 percent raise each year, and your savings earn 8 percent, you will have $1 million in 30 years—just in time for retirement.[2]

Don't forget about inflation . . .

Sorry to be a downer, but most savers forget that a penny saved today is not a penny earned 20 years from now. Let's look at the example where you were going to save $3,000 a month and earn $1 million in 15 years. When you take 3.1 percent annual inflation into account, your money would really be worth $635,876 in 15 years. *Ay, caramba!* (If you want to play with the numbers, go to Bankrate.com and search for "Save a million dollars calculator.")

How long does it take to count to 1 million?

About 21 days. If you are able to count from 1 to 100 in one minute—a rate of about 6,000 counts per hour—you could reach one million in 166.7 hours. If you counted for 8 hours each day, you'd be done in approximately 21 days.

Time Trivia: Credit report times

They might seem boring (okay, they *are* boring), but credit reports can make or break a loan deal; they can even influence whether a potential employer decides to hire you. Keep yours in good shape.

How long do mistakes stay on your credit report?

Negative information, like late payments, stays on your report for 7 years; bankruptcy will remain for 10.

How often can you request a free copy of your credit report?

Every 12 months. According to the Federal Trade Commission, consumers are allowed to order one *free* copy of their report from each of the national consumer reporting companies each year. (You can view your reports online instantaneously at www.annualcreditreport.com.)

Education & Money

Is a college degree still worth the time and money investment? Yes, with some caveats. A typical college graduate earns $650,000 *more* than a typical high school graduate over the course of a 40-year career, according to the Pew Research Center. That's a lot of dough. This figure takes into account the up-front costs of tuition and related college costs.

Here's another way to look at the benefit of a college degree: in 2008, workers with bachelor's degrees earned 65 percent more than high school graduates ($55,700 vs. $33,800). Workers with associate's degrees earned 73 percent more than those who had not completed high school ($42,000 vs. $24,300).[3]

However, there are exceptions. For instance, 27 percent of people with postsecondary licenses or certificates—such as paralegals or police officers—earn more than the average bachelor's degree recipient.[4]

Some degrees are worth more than others

Not all college degrees are created equal. Students who receive a degree from MIT will earn $1.355 million *more* over a 30-year career than students with just high school degree. On the other hand, artistes who attend Savannah College of Art & Design will earn $160,000 *less* than a high school grad over a 30-year work span (presumably because of the cost of the degree).

Public school degrees also vary widely in the earning power they bestow on grads. Students from the University of California at Berkeley, one of the country's premiere state schools, will earn $836,200 more over 30 years than a high school grad, but students with a degree from Concord University in West Virginia will earn $137,000 less than a high school grad, according to calculations by *Bloomberg Businessweek*.[5]

How long will it take to recoup your MBA tuition?

From 2 to 8 years, according to *Bloomberg Businessweek*. The magazine took the top-rated business schools from their 2010 rankings and then, based on the tuition the schools charged and the salaries their grads earned, figured out how long it would take for grads to break even.

If you went to SDA Bocconi in Italy, it would take just 2.25 years for you to make your MBA investment back. (I'll bet that number doesn't factor in travel costs between Italy and the United States.) If you stayed closer to home and went to Michigan, the school at the bottom of *BBW*'s list, it would take nearly 8 years.[6]

Here's a sampling from the rankings, including the school with the fastest return on investment and the longest.

Business School	Years to recoup costs	BBW 2010 Best Business School Rank
SDA Bocconi	2.25 years	18
HEC Paris	2.3	14
Penn State (Smeal)	3.98	44
Wisconsin-Madison	5.42	34
INSEAD	5.59	1
Boston University	6.15	38
Duke (Fuqua)	6.5	6
Yale	6.8	21
Dartmouth (Tuck)	7.13	14
Michigan (Ross)	7.48	7

Tax Time

How long does it take to complete federal income tax returns?

It takes an average of 18 hours to put together the federal forms, according to the IRS. But 68 percent of taxpayers use Form 1040, and that form takes a whopping 22 hours to complete, from paper collection to submission.

The Form	Total Time
Form 1040	*22 hours:* 10 hours for record keeping, 3 for tax planning, 5 for form completion, 1 for submission, and 3 for "other."
Form 1040A	*10 hours:* 4 for record keeping, 1 for tax planning, 3 for form completion, 1 for submission, and 1 for "other."
Form 1040EZ	*7 hours:* 2 for record keeping, 1 for tax planning, 2 for form completion, 1 for submission, and 1 for "other."

Time Trivia: Men are twice as likely to wait until tax day to file their returns than women: 3 percent of men wait till T-Day to file versus just 1 percent of women who do.

Which Do You Think About More Often, Money or Time?

Think about the question. Now consider what your answer may mean:

In two different experiments conducted at the Wharton School, people were primed to focus on either time or money. When the participants focused on time, they were more likely to engage in *social activities* during the experiment or to say they would do social things the next day. But when individuals were primed to think about money, they spent more of their time *working* and less of their time with pals, during the experiment or after.

"Focusing on money motivates one to work more, which is useful to know when struggling to put in that extra hour of work to meet a looming deadline," wrote Cassie Mogilner, author of the study "The Pursuit of Happiness: Time, Money, and Social Connection."[7] "However, passing the hours working (although productive) does not translate into greater happiness. Spending time with loved ones does, and a shift in attention toward time proves an effective means to motivate this social connection."

In short, if you want to be happier, think about time.

If you want to be more productive, think about money.

Grooming time and your paycheck

What does primping have to do with your paycheck? Depends on whether you're a man or a woman, black or white. One study[8] (which I also mentioned in Chapter 1) found that the more time women spent primping, the less money they made in weekly earnings. It's hard to say for certain which is the cause and which is the effect, but the authors theorize that the "reason may have to do with the negative stereotypes associated with an 'overly groomed' woman in the workplace."

Men's habits had a more complex effect on their earnings. Grooming had no effect on white men's earnings, but for minority men increased grooming time had "an unambiguously positive and large effect on their earnings." Why? "While we cannot offer a definite explanation at this point, it may be that grooming helps to counter negative stereotypes regarding minority men's agreeableness or conscientiousness," the authors wrote.

Workout time and your paycheck

Want to earn more? Then work out more. Employees who regularly exercise earn 7 to 11 percent more than those who don't exercise, according to a study conducted by Vasilios D. Kosteas of Cleveland State University. Kosteas found, for example, that men who exercise 3 or more times a week earn 6.6 percent more than those who do not, while women who exercise similarly earn 11.2 percent more. The author concludes, "while even *moderate* exercise yields a positive earnings effect, *frequent* exercise generates an even larger impact."[9] Earlier studies have found that obese individuals tend to earn less than people of normal weight.

What's the best *age* at which to make financial decisions?

Age 53. Experience rises with age, while analytical abilities decline. At this cross section—when wisdom is ripe and critical thinking is still sound—is the auspicious age of 53, according to research by a team of four economists. The authors of "The Age of Reason" looked at studies on cognitive function as well as the borrowing habits of adults. The researchers found that the ability to understand complex variables begins to drop off at age 20. (Don't panic, the decline is very slow.) Real impairment does not start until much later. The researchers also looked at a cross section of prime borrowers and found that middle-aged people were wisest, borrowing at lower interest rates and paying fewer fees relative to younger and older adults.[10]

How long do recessions last?

A recession is a period of diminishing economic activity that lasts on average, in the United States, for 17.5 months. Recessions are named and measured retroactively by the National Bureau of Economic Research. The NBER, for instance, determined the last recession's start date of December 2007, 11 months *after* that date; they announced the end date of June 2009 in September 2010.

Between 1854 and 2009 there have been 33 recessions. The longest recession on record was from October 1873 to March 1879—65 months!—and is often referred to as the Long Depression. The second-longest recession occurred during the Great Depression, August 1929 to March 1933—43 months.

The recession we just experienced was the longest since World War II.[11]

A Summation of the Last Four Recessions	
December 2007 to June 2009	18 months
March 2001 to November 2001	8 months
July 1990 to March 1991	8 months
July 1981 to November 1982	16 months

What is your time worth?

Ian Walker, a British economist, came up with a formula that determines how much your time is really worth. The formula takes into account wages, taxes, and the local cost of living. It goes like this:

$$V = \frac{[W(100\text{-}T)]}{C}$$

W is your hourly wage

T is your federal tax rate

C is the local cost of living

V is your hour's value

Let's say you earn $100 an hour, are in the 33 percent tax bracket, and the cost of living in your area is 115 (the index for Chicago at press time. You can find the index in your area by doing an Internet search).

Subtract your federal tax bracket from 100: 100 – 33 = 67

Multiply your hourly wage by that number: 67 * 100 = 6700

Divide the result the by average cost of living for your area: 6700 / 115 = $58

In this example, one hour of your time is worth $58.

A simpler way to find your real hourly rate is to use the "Time Is Money" calculator at SmartMoney.com. The calculator is based on Walker's formula and doesn't ask for your cost of living index—just whether you live in a suburban, urban, or superurban area.

Use your hourly rate to determine whether you should hire someone to do projects for you—and how much money you waste when you take a nap.

What is a stay-at-home mom's time worth?

Six figures. The typical stay-at-home mom works almost 95 hours a week, according to a 2012 survey by Salary.com. (About 37 percent of moms are considered stay-at-homes, according to a Gallup survey.[12]) The folks at Salary.com evaluated the schedules of 8,000 mothers and divvied their work into 10 different duties. According to the site's calculations, women deserve to be paid an annual salary of $112,962 for their work. Here's how they came up with that number: stay-at-home moms work a total of 94.7 hours a week, 40 hours base with 54.7 hours of overtime. That comes to an annual salary of $112,962, or $17.80 an hour. Of course, this calculation is gimmicky—and some moms might find it insulting. But if you don't take it too seriously, it's interesting to see how statisticians translate a mother's work into discrete job categories.[13]

Weekly hours

14.8 hours per week as a housekeeper

13.9 hours as a chef

13.7 hours as a day-care teacher

10.7 hours as a facilities manager

8.9 hours as a PC or Mac operator

7.9 hours van driver

7.7 hours as a janitor

7.6 hours as a psychologist

6.3 hours doing laundry

3.2 hours as household CEO

Q&A: When it comes to money, forget the past

Huh? That doesn't sound like smart advice at all. After all, aren't we supposed to learn from the past? Bear with me. To help you understand why our past actions can sometimes have a problematic effect on our current decisions, I spoke to Gary Belsky, award-winning writer and coauthor of *Why Smart People Make Big Money Mistakes and How to Correct Them: Lessons from the Life-Changing Science of Behavioral Economics.*

LA: What bearing do our past actions have on our present decisions about money?

GB: We pay far too much attention to them! In psychology they call this the "sunk-cost fallacy." The idea is that we let previous investments influence current decisions, because we don't want to "give up" on what we've already spent. It's why people pour good money after bad, in general, but also good time after bad; i.e., it's the reason people often stay in bad relationships or lousy movies, and why they can't stop reading a boring book once they're into it. They say to themselves, "Well, I've already spent this much time . . ." My advice: Forget the past, since it's already happened.

LA: Why is it so hard to let go of poorly performing investments?

GB: Well, the sunk-cost fallacy is one reason. But some other forces come into effect. One is called "loss aversion." The idea is that we're especially sensitive to losses, more than we should be. As a result, we have a hard time selling losing investments because we're always secretly hoping they'll bounce back, at least to the price we paid for them. And that price becomes way too important to us, because of something called anchoring. Anchoring is basically the idea that because life is so complex, and because we have to make so many decisions, we tend to arbitrarily pick anchors to start our decision-making process with. That's fine, but too often the anchors become too important.

LA: When is it OK to sell something (an investment, a house) at a loss?

GB: It doesn't matter what you paid for an investment. What matters is what it's worth now, and what it might be worth in the future. That's the only consideration. Sometimes I tell people to ask themselves a different question, which is "Would you *buy* the investment again tomorrow?" If the answer is no, then you have to seriously think of selling it. Think of it this way: if I bought an investment for $100 five years ago that's only worth $50 today, every day that I *don't* sell it is really a decision to put that $50 back into it. When you think about it on those terms, it becomes a lot easier to get rid of losing investments.

Time and attention are two things, like money,
that we tend not to notice until they are in short supply.
—Merlin Mann, founder of 43Folders.com

 Time Exercise 8: Since money won't make you happy, try time

Over the past decade researchers have learned that having lots of money increases your happiness by only a small percentage. Gaining more time, on the other hand, can increase your happiness a lot, provided that you use your time well.

What does *using time well* mean? Three scholars answered that question in a recent article, "If Money Doesn't Make You Happy, Consider Time," which appeared in the *Journal of Consumer Psychology*.[14] The authors provide these 5 suggestions for wise time use:

1. **Spend time with the right people**. Be honest. How many people in your life irritate you; and how much of your precious time do these folks consume? Ditch the whiners and complainers, and spend more time with friends who make you feel good. This goes for your work life too.

Find people on the job who make you as happy as friends off the job. "Two of the biggest predictors of people's general happiness are whether they have a 'best friend' at work and whether they like their boss," the authors note. Not much you can do about the boss, but you can certainly work on cultivating a BFAW (best friend at work).

2. **Spend time on the right activities.** Be honest, again. What makes you *really* happy? One way to evaluate options about how to spend your time is to consider whether the "experience will remain sticky over the long run," say the authors. "Sticky experiences are more valuable because they temporally extend the pleasure of a single moment." Go shopping or take a hike in a new location—which seems stickier to you?

3. **Spend time in contemplation.** This is my favorite tip. You don't have to *do* something to *enjoy* it. Just contemplating a vacation or a dinner with your buddies can make you feel really good. In Chapter 3 (page 86), I mention a study that found that that the happiest part of a vacation is the planning part. "Research suggests that we might be just as well off, or even better off, if we imagine experiences, but not have them. So, spend plenty of time happily daydreaming," the authors conclude.

4. **Expand your time.** If you feel constantly time pressed, consider buying time—literally: hire a babysitter, a cleaning person, a dog walker. If that's not an option, then be here now. Being present (a.k.a. not distracted) slows the perceived passage of time, allowing you to feel less rushed, hurried, and harried.

5. **Don't plan your future based on your present attitudes about happiness.** Happiness is dynamic over time. Younger people, for example, are more likely to associate happiness with excitement, whereas older individuals are more likely to experience happiness as feeling peaceful. Be cautious about basing decisions about your future happiness on your current perceptions of bliss.

Media

Gadgets, the Internet & Multitasking

All things digital—e-mail, social media, online games, and the infinite allure of the web—claim a skyrocketing amount of our time. Some of that time is productive, while some is—of course—pure fun and folly. These media are always available and always on, thanks to iPhones, BlackBerrys, iPads, PCs, and others.

Kids today have become multitasking maestros, spending much of their day engrossed in a few forms of media at once. Adults are catching up. Nearly half of tablet owners use them daily while watching TV; workers keep an average of 8 windows open on their computers at once. Why do we all keep multitasking when we know it makes us less productive? This chapter has some of the answers.

The one thing that's predictable is that by the time this book is out, the tech-startup generation will have created hundreds of new ways to get our attention—and one of them may be radically influencing our lives.

Internet Time: The Big Picture

It's mind-boggling: just over 15 years ago the Internet as we know it didn't exist. Now we're hooked. In 2012, Americans spent 30 hours a month online, up 10 percent from a year earlier—and that's just an average of the entire population. (I easily spend 30 hours online each *week,* doing research, working, and shopping.)

Meanwhile, the number of Americans with access to the Internet has more than doubled over the past decade, growing from 132 to 274 million.

Internet time

What do people do when they are online? They spend most of their time on social networking sites and blogs. Nielsen, the media research company, came up with this breakdown, which shows how Internet jockeys allocated their time to various activities in 2011.

Category	Percentage of time spent
Social networks and blogs	21.3
Games	7.7
E-mail	6.5
Videos/movies	4.3
Portals	3.8
Other	56.5

Mobile Internet time

Those with Internet-enabled phones spend the majority of their time using apps. Data, again, are courtesy of Nielsen:

Function	Percentage of time spent
Text messaging	13.4
Browser	11.1
Social network apps	5.5
Dialer	5.4
E-mail	5.3
Music/video apps	2.3
Camera	1.1
Other apps	55.8

Which sites do people spend the most time on?

Facebook! Users spend 6-plus hours a month on the site, followed by AOL Media Network (2-plus hours), and then Yahoo! (2-plus hours).

The defining and ongoing innovations of this age—
biotechnology, the computer, the Internet—
give us a chance we've never had before to end
extreme poverty and end death from preventable disease.
—Bill Gates, Harvard commencement speech

E-mail Time

Got mail? The vast majority of e-mail users (87%) check their personal accounts more than once a day; 29 percent check more than 4 times a day.

Most of us can't turn off. According to a survey by Harris Interactive for Xobni, a software company, 72 percent of Americans regularly check their e-mail on vacations and weekends; 20 percent of professionals who check e-mail after hours do so in bed—either as the first thing they do in the morning or before going to sleep at night.

Can *you* resist a message?

They're so enticing. Maybe the message has hot gossip, or an invitation, or a piece of information you desperately need. A 2010 survey asked adults: "In which situations will you stop what you are doing to read an electronic message?" The surveyors divided the responders into two groups, those under 25 and those over. The 25+ group had better boundaries—62 percent of respondents in the older group said flat out: "I don't like interruptions." Here's how the two groups answered.[1]

In which situations will you stop and read an electronic message?

	Percentage who said yes	
	Under age 25	**Over age 25**
In a meeting?	22	11
Eating a meal?	49	27
Having sex?	11	6
On the potty?	24	12

 Time Trivia: How old is e-mail?

How did we live without it? E-mail became official on August 30, 1982. That's the day then 16-year-old V. A. Shiva Ayyadurai copyrighted the term "EMAIL." Dr. Shiva, also known as Dr. Email, now has four degrees from MIT. His resume boasts seventeen different inventions; he is also a lecturer at the Biological Engineering Division of MIT.

Time Exercise 9: Create a distraction-free hour

Whether you work in a busy office or at home alone, you need time that's free from interruptions to create, think, and focus. Schedule one hour each day for uninterrupted work; don't answer the phone, don't look at e-mails, don't talk to colleagues. Ideally this hour is the same each day, but for some workers this may be impossible. If this idea makes you anxious, then all the more reason to try it. Start a revolution in your office: suggest that one hour of the workday be meeting, email and phone free.

Time Trivia: The tyranny of the in-box

One-third of all people would rather clean their toilets than clean out their e-mail in-box, according to a Yahoo! survey.

Social Networking Time

Americans spend many hours each month keeping up with their pals on social networking sites. The number of folks using social networking sites has nearly doubled since 2008, and the population of users has gotten older.[2] More than one-third of Americans, 36 percent, admit that social networking is their biggest time waster, followed by fantasy sports at 25 percent, and TV at 23 percent.[3]

The favorite site? Facebook, of course.

Some Quirky Stats About FB

- The average FB user has 229 friends; the younger the user, the more friends he or she has.

- FB'ers make 7 new friends each month.

- Women update their status 21 times each month, on average, while men update just 6 times.

- 18 percent of youngsters on FB (18–22 years old) update their status daily; 13 percent update several times a day.

Who spends the most time socializing online?

Israelis. According to comScore, an Internet analytics company, Israeli visitors spent 10.7 hours a month on social networking sites in 2011.[4]

Country	Hours per visitor/month
Israel	10.7
Russia	10.3
Argentina	8.4
Philippines	7.9
Turkey	7.8
Venezuela	7.0

Colombia	7.0
Chile	6.7
Canada	6.4
Spain	5.7
United Kingdom	5.3
Mexico	5.3
United States	5.2
Malaysia	5.1
Germany	5.0
Puerto Rico	4.9
Peru	4.9
Italy	4.8
Indonesia	4.6
Norway	4.5

Time Trivia: Networking in the Loo

Nearly a third of 18–24-year-olds use social networking in the bathroom.

Cell-Phone Time

Cell-phone owners spend most of their time talking, texting, and e-mailing. Smart-phone owners (half of Americans own one), however, spend two-thirds of their time using apps.

Calls are getting shorter and less frequent as more people prefer to text rather than talk. In 1991, calls lasted 2.38 minutes, 20 years later an average local call lasted just 1.78 minutes. FYI: the average local monthly wireless bill (which includes voice and data service) was $47.00 in 2011, down from $72.74 in 1991.

Women talk and text more than men

Women chat on their cell phones for 770 minutes per month, while men chat for just 600 or so minutes. Women win on texting too, sending or receiving 800+ SMSs per month, compared to men who send/receive around 600.

Some cell-phone users find that their phones have added benefits:[5]

- 42 percent of cell owners use their phone as entertainment when they are bored.

- 13 percent pretend to be using their phone to avoid interacting with others.

TV Time

TV still absorbs more of our downtime hours than other electronic temptations. But most of the time, we are doing something else as well—using a tablet, checking the Internet, texting friends.[6]

Who spends the most time watching TV?

Older folks. Americans 65 years and older spend 204 hours per month watching the telly, compared to 145 hours for the average American.

African Americans spend the most time watching TV (206 hours), followed by whites (142), Hispanics (125), and Asians (96).

Americans love TV so much, they have multiple sets. The average household has 2.5 TVs; one-third of households have 4 or more.

Age	Monthly TV Time (hours)
2 to 11	116
12 to 17	109
18 to 24	113
25 to 34	125
35 to 49	143
50 to 64	178
65+	204

How much time do we spend watching videos & TV each month?

Watching TV	144.5 hours on average
Watching video on the Internet	15.5 hours
Watching video on mobile phone	15.2 hours

What do we watch during prime time?

Dramatic shows captivate *more* attention than reality shows or sports. Television prime time runs from 8 P.M. to 11 P.M. on Monday through Saturday, but Americans tune in to shows most often between 9:15 P.M. and 9:30 P.M. Sunday night PT is 7 P.M. to 11 P.M.[7]

Share of prime-time viewing, men vs. women[8]

	Men	Women
Drama	34.5%	46.9%
News	9.3	9.9
Reality	12.9	17.8
Sitcom	10.7	12.0
Sports	32.7	13.5

Game Time

The average gamer spent 13 hours playing each week in 2010. But a small cadre of devotees (4%) spent a whopping 48.5 hours per week engaged with their games.[9] Games are not just for kids. Nearly a third, 29 percent, of gamers are over 50; just 18 percent are kids.[10]

Games People Play

Top 5 PC Games	Minutes played per week
World of Warcraft	500
League of Legends	402
Hanging Gardens of Babylon	267
Lord of the Rings	527
Half Life 2	265

News Time

The Internet has surpassed newspapers as the go-to news source. Nearly half of Americans (46%) say they get news online at least three times a week, compared to just 40 percent who rely on newspapers on a weekly basis, according to the Pew Research Center's Project for Excellence in Journalism.[11] The typical American spends time with a variety of news media, spending 70 minutes watching, reading, and listening to news on any given day.

Daily news time, 2004 vs. 2010[12]

	2004	**2010**
Watching TV news	32 minutes	32 minutes
Listening to radio news	17 minutes	↓15 minutes
Reading news online	6 minutes	↑13 minutes
Reading the newspaper	17 minutes	↓10 minutes

How much time do people spend reading magazines, online and off?

Technology has been great for magazines. Readers spend more time with their issues when they read them on an e-reading device than when they read them the old-fashioned way.

Time spent reading a magazine . . .

If you're reading it on an iPad or tablet	60 to 80 minutes per issue
If you're reading a hardcopy	45 minutes per issue
If you're reading it on a computer	30 to 40 minutes per issue

Time Trivia: Presidential Sound Bites

In 1968, presidential sound bites on the evening news lasted 42 minutes; by 1988, that number had dropped down to 9.8 minutes. In 2000, it was 7.8 minutes (the last year for which there is data).

Wired Kids

Children use an astounding array of digital media, from smart phones, to handheld video-game players, to tablets. The favored medium is still television: most kids watch about 3 hours a day. But by age 4, 20 percent of children are playing handheld video games; by age 7, nearly half of all kids are navigating games on handheld devices. Multitasking starts early. Among children ages 2 to 11, more than one-third use the TV and Internet simultaneously.

All this media consumption led tech guru Don Tapscott to remark in *AdAge*: "This is the first time in human history when children are an authority on something important." Namely, media. "Today, the 11-year-old is an authority on this digital revolution, which is changing business, commerce, government, entertainment—every institution in society."[13]

Older kids (8 to 18) spent 7 hours and 38 minutes a day using entertainment media in 2010, according to a comprehensive survey by the Kaiser Family Foundation. During that time, two-thirds of kids were media multitasking—as they do so well—listening to music, texting, and talking on their phones all at once. Kaiser estimates that all this multitasking means that kids actually pack 10 hours and 45 minutes of media into those 7.5 hours. Imagine how these numbers will change if researchers revisit this topic in 2013?[14]

How much time do kids spend with various media?

About 30 percent of kids use more than one form of media at once, which is why there are two total lines below:

	8- to 10-year-olds	11- to 14-year-olds	15- to 18-year-olds
TV	3 hrs., 41 mins.	5 hrs.	4 hrs., 22 mins.
Music	1 hour, 8 mins.	2 hrs., 22 mins.	3 hrs.
Computers	46 mins.	1 hr., 46 mins.	1 hr., 39 mins.
Video games	1 hr.	1 hr., 25 mins.	1 hr., 8 mins.
Print	46 mins.	37 mins.	33 mins.
Movies	28 mins.	26 mins.	20 mins.
Total media use:	5 hrs., 29 mins.	8 hrs., 40 mins.	7 hrs., 58 mins.
Total media exposure: (includes multitasking)	7 hrs., 51 mins.	11 hrs., 53 mins.	11 hrs., 23 mins.

Time with media varies by race and income

Black and Hispanic youths (ages 8 to 14) spend about 13 hours a day using media; whites spend about 8.5 hours, according to the Kaiser Family Foundation.[15] In general, kids in lower-income homes spend more time with technology.

Teens and Texting

Talk about addicted. About 77 percent of teens own a cell phone; 23 percent own a smart phone. Middle and high school kids spend an average of 1 hour and 35 minutes a day sending or receiving texts. Teens send a median of 60 texts a day, up from 50 in 2009. Girls text more than boys: older girls are enthusiastic texters, sending a median of 100 texts, compared with 50 for boys the same age. A hypersocial 18 percent of teens send and receive more than 200 texts a day![16,17]

Texting on the road

Yikes: 43 percent of teen drivers admit to texting while driving; 60 percent have texted at a red light; and 73 percent have glanced at their phone at a stop light.[18]

Time Trivia: Teen Texters

The majority of teens expect a reply to a text or an e-mail within five minutes or less.[19]

Shopping Time

I love to shop online. According to a survey by Power Reviews, so do a lot of other people. Fully 30 percent of online shoppers shop weekly; 42 percent shop several times a month. Half of shoppers devote the majority of their shopping time to researching products.[20]

Which retailers take the time to respond to your questions?

To determine which vendors take the time to answer customers' questions in full, STELLAService sent daily emails to 25 top notch retailers with queries such as, "I live in an apartment and I always miss my delivery guys when they come. How many times will they attempt?" STELLA peppered the retailers with questions for 45 days. The most thorough answers came from:[21]

LL Bean. The Maine-based retailer answered 88.9 percent of the emails completely over the 45-day study

Gap.com responded to 84 percent of emails completely

Zappos.com, 75 percent

VictoriasSecret.com, 70.4 percent

TigerDirect.com, 70.4 percent

How long are customers willing to wait on hold?

Surprisingly, consumers say they are willing to wait, on average, for up to 13 minutes for a customer service representative to help them out.[22]

Time on hold	Percent who are willing to wait
Less than 5 minutes	19
5 to 10 minutes	28
10 to 15 minutes	21
15 to 30 minutes	21
30 minutes to 1 hour	5
More than an hour (!)	7

Multitasking: deliciously addictive

We all now know that multitasking is inefficient—but most of us do it anyway.

- More than one-third of tablet owners report using their devices while in the bathroom!

- American workers keep an average of 8 windows open on their computer screens at once and toggle between them every 20 seconds.

- Half of adults use their tablet and TV simultaneously daily.

A study conducted at the Brain, Cognition and Action Laboratory at the University of Michigan found that when individuals switched back and forth between two projects, that is, multitasked, it took longer for them to complete the tasks than when they did the same tasks sequentially. "For some projects, it took at least 50 percent longer to finish each one when multitasking was involved," Dr. Meyer, director of the laboratory, explained to me in an e-mail.

Why do we keep doing it? Because it feels good. A recent Ohio State University study followed a group of multitasking college students and found that multitasking gave the students an emotional boost, even when it interfered with their main task, such as studying. Students who watched TV while reading a book, for instance, reported feeling more emotionally satisfied than those who studied without watching TV, but they also admitted that they didn't achieve their goals as well. Researchers also found that students became accustomed to multitasking and the benefits they derived from it, which made them more likely to keep at it. Multitasking, like other habits that make us feel good (eating chocolate, watching sports, drinking martinis), can be addictive.[23]

We may also keep multitasking because we hope to get better at it. There's some evidence that multitasking ability improves with practice. A very small portion of the population, about 2.5 percent, are actually expert multitaskers. Are you one of them?

Why are we so distractible?

Because we are hardwired with intense and immediate needs. We feel thirsty, hungry, amorous, or lonely. These desires bombard us all day long, distracting us from the task at hand. In a recent study,[24] researchers gave 205 people BlackBerrys that were programmed to go off at random intervals. When the devices went off, users were asked whether they were experiencing, or had recently experienced, a desire. The results? Participants felt some form of desire about half the time they were awake; what's more, they spent 3 to 4 hours trying to resist those desires, according to Roy Baumeister, one of the study's authors. The most common desires were the need to eat, sleep, or drink—followed by the desire for media use and social contact. Here's the problem: almost half of these desires were described as conflicting with the person's other goals, values, or motivations.

Individuals tried different ways to resist, reporting that it was easiest to ignore the desire to sleep, have sex, and spend money, and hardest to resist the desire to watch TV, surf the web, or just tune out and relax. Some people were more successful than others at resisting desire. "People are not helpless in the face of their motivations but instead can adjudicate whether to enact them or not," the authors wrote. But adjudication is a tricky process and many issues come into play. "Motivations compete with each other and with various inner processes and external realities to drive behavior. The path from human desire to behavior is apparently rarely a simple and straight one." I'll say!

There is more to life than increasing its speed.
—Mahatma Gandhi

Is Tech Time Ruining Our Lives, Our Culture, Our World?

Each month there's a dire new headline: "Is Facebook Making Us Lonely?" "Can Too Much Texting Make Teens Shallow?" "Will Hyper-Connected Millennials Suffer Cognitive Consequences?"

These are valid questions. Media has a potent allure that often overrides our better judgment. (It's not called the "Crackberry" for nothing.) Checking e-mail, surfing the web, and engaging with social networks give us instant feedback—which can be addictive. What's more, society now expects us to be always connected, so there's social pressure to be always on, always available.

There's cause for worry, but there's also cause for celebration. Playing video games has been shown to help brains develop and stay sharp. Adults with e-readers say the devices encourage them to read more. Experts surveyed for a recent study, the *Future of the Internet,* said the effects of hyperconnectivity and the always-on lifestyles of young people will be mostly positive. Yes, these kids will grow up with a desire for instant gratification, but they will also be quick-witted analysts and savvy decision makers.[25]

And consider the big picture. Steve Jobs pointed out in his 2005 address to Stanford graduates that technology has "brought the world a lot closer together, and will continue to do that. There are downsides to everything; there are unintended consequences to everything. The most corrosive piece of technology that I've ever seen is called television—but then, again, television, at its best, is magnificent."

> *Observe a method in the distribution of your time.*
> *Every hour will then know its proper employment,*
> *and no time will be lost.*
> —George Horne, English bishop

**Longevity
Humans & Animals**

How long will you live? The answer depends on many factors: where you live, what you eat, how much you exercise, and even a lick of luck. A person born in the United States today is projected to live for 78 years. But if that person does most of the *right* things and few of the *wrong* ones, and doesn't spend too much time in automobiles (a dangerous place to spend time), he could easily live 10 years longer. You have more control over your longevity than you might think.

If thinking about aging is disheartening, then skip straight to page 317. It's not all downhill. Three experts point out that there are, in fact, some benefits to growing old.

Life Span

How long do Earthlings live? A global comparison

First, some benchmarks: the average global life expectancy is 67.07; life expectancy in the USA is 78.49. Where do people live the longest? Monaco! The lucky Monégasques live, on average, until they are 90 years old, the longest life expectancy on the planet. (Might this be because Monaco boasts the highest GDP per capita in the world, or is it all that gambling that keeps its citizens young?) Here are the top 50 life expectancies, plus the bottom 10, according to the *CIA's World Factbook*.

Ask yourself this: Why does the USA come in at number 50?[1]

Life Expectancy at Birth

Rank	Country	Years
1	Monaco	89.68
2	Macau	84.43
3	Japan	83.91
4	Singapore	83.75
5	San Marino	83.07
6	Andorra	82.50
7	Guernsey	82.24

Rank	Country	Years
8	Hong Kong	82.12
9	Australia	81.90
10	Italy	81.86
11	Liechtenstein	81.50
12	Canada	81.48
13	Jersey	81.47
14	France	81.46
15	Spain	81.27
16	Sweden	81.18
17	Switzerland	81.17
18	Israel	81.07
19	Iceland	81.00
20	Anguilla	80.98
21	Netherlands	80.91
22	Bermuda	80.82
23	Cayman Islands	80.80
24	Isle of Man	80.76
25	New Zealand	80.71

26	Ireland	80.32
27	Norway	80.32
28	Greece	80.31
29	United Kingdom	80.30
30	Germany	80.19
31	Jordan	80.18
32	Saint Pierre and Miquelon	80.00
33	Austria	79.91
34	Faroe Islands	79.85
35	Malta	79.85
36	European Union	79.76
37	Luxembourg	79.75
38	Belgium	79.65
39	Virgin Islands	79.47
40	Finland	79.41
41	South Korea	79.30
42	Turks and Caicos Islands	79.26
43	Wallis and Futuna	79.12
44	Puerto Rico	79.07

Rank	Country	Years
45	Bosnia and Herzegovina	78.96
46	Saint Helena, Ascension, and Tristan de Cunha	78.91
47	Gibraltar	78.83
48	Denmark	78.78
49	Portugal	78.70
50	United States	78.49
****************	******************	****************
212	Mozambique	52.02
213	Lesotho	51.86
214	Zimbabwe	51.82
215	Somalia	50.80
216	Central African Republic	50.48
217	Afghanistan	49.72
218	Swaziland	49.42
219	South Africa	49.41
220	Guinea-Bissau	49.11
221	Chad	48.69

How long do Americans live—a cross-country lineup

If you'd like to live a long, healthy life in the United States, move to Hawaii. Aloha! The youngest state in the nation has the highest life expectancy. Perhaps it's because the island state has low levels of unemployment and poverty, and a health index of 6.45, well above the national average of 5.25, according to the American Human Development Project.[2]

State	Life expectancy
Hawaii	81.5
Minnesota	80.9
California	80.4
New York	80.4
Connecticut	80.2
Massachusetts	80.1
North Dakota	80.1
Utah	80.1
Arizona	79.9
Colorado	79.9
South Dakota	79.9
Florida	79.7
New Hampshire	79.7

Vermont	79.7
Iowa	79.7
New Jersey	79.7
Washington	79.7
Wisconsin	79.3
Rhode Island	79.3
Idaho	79.2
Nebraska	79.2
Oregon	79.0
Illinois	78.8
Maine	78.7
Virginia	78.5
Kansas	78.4
Montana	78.4
Delaware	78.3
Alaska	78.3
Texas	78.3
New Mexico	78.2
Pennsylvania	78.2

Maryland	78.1
Michigan	77.9
Indiana	77.7
Wyoming	77.6
Nevada	77.6
Ohio	77.5
Missouri	77.4
North Carolina	77.2
Georgia	77.1
South Carolina	76.6
Tennessee	76.2
Kentucky	76.2
Arkansas	76.1
District of Columbia	75.6
Oklahoma	75.6
Louisiana	75.4
Alabama	75.2
West Virginia	75.2
Mississippi	74.8

Time Exercise 10: How long might _you_ live?

To get a rough estimate of your projected life span, use the government's _Life Expectancy Calculator_ at socialsecurity .gov.[3] To get a precise estimate, try the _Life Calculator_, devised by experts from the Wharton School at the University of Pennsylvania (Google: "life calculator + Wharton"). The _Life Calculator_ takes about 5 minutes to complete and asks very detailed questions about your daily habits, the history of cancer in your family, how much you travel, and more.[4]

How to Live Longer

If you follow the advice you've heard ad infinitum—eating well, exercising often, staying slim, and abstaining from cigarettes and other noxious substances—you can add years to your life. Really.

Lifestyle has a big impact on life span. A Swedish study followed 855 men born in 1913 for more than 50 years and found that lifestyle had the biggest impact on the participants' longevity, more so than heredity. Men who did not smoke, were in a good socioeconomic bracket, consumed moderate amounts of coffee, had low cholesterol, and were still working at 54 had the greatest chance of celebrating their 90th birthday.[5]

To lengthen your life span, DO:

Eat like a Mediterranean

That means consuming lots of vegetables, fruits, fish, and olive oil—which are naturally low in calories and unhealthy fats, and high in antioxidants and fiber. The so-called Mediterranean diet has been shown to lower the risk of heart disease, high blood pressure, stroke, diabetes, and breast cancer—conditions that shorten one's life span. A report published in 2000 looked at three major studies on this particular way of eating and concluded, "a diet that adheres to the principles of the traditional Mediterranean one is associated with longer survival."[6]

Another recent study showed that eating fiber keeps you young. Researchers followed 388,000 people (ages 50 to 71) over 9 years and found that those who consumed the most fiber were 22 percent less likely to die from any cause than those who ate the least. (Men should consume 38 grams of fiber a day, women 25.)[7]

Skeptical? Here's another reason to eat like a Greek, or an Italian: adults who adhere to a Mediterranean-style diet have less cognitive decline in old age than those who eat any old way.

Exercise—even a little goes a long way

The current recommendation is to exercise for 30 minutes most days of the week. But (drum roll, please), working out for just 15 minutes per day can add three years to your life. A study published in *The Lancet* in 2011 followed 416,175 people for 12 years and found that those who exercised for an average of 92 minutes per week, or 15 minutes a day, had a 3-year longer life expectancy than those who didn't exercise at all. In addition, "Every additional 15 minutes of daily exercise beyond the minimum amount of 15 minutes a day further reduced all-cause mortality by 4% and all-cancer mortality by 1%. These benefits were applicable to all age groups and both sexes, and to those with cardiovascular disease risks. Individuals who were inactive had a 17% increased risk of mortality compared with individuals in the low-volume [15 minutes a day] group," the study's authors wrote.[8]

Be conscientious

People who are prudent, persistent, and well organized seem to live longer than those who are not. This makes intuitive sense. If you are conscientious, you're more likely to avoid destructive behaviors (like drugs and drunk driving) and adopt healthy ones (exercise, regular checkups). *The Longevity Project,* a book that analyzed data from the decades-long Terman Study of the Gifted, found that conscientiousness was one of the strongest predictors of longevity. The Terman study followed 1,500 gifted Californians born in 1910 for most of their lives. The book's authors found that "by the end of the twentieth century, 70 percent of the Terman men and 50 percent of the women had died. . . . It was the unconscientious among them who had been dying in especially large numbers."[9]

Work in a friendly office

Workers who have supportive colleagues are healthier and longer lived than workers who toil in an unpleasant environment, according to an Israeli study. Researchers at Tel Aviv University followed the health records of 820 adults, ages 25 to 65, who worked an average of 8.8 hours a day over two decades. Those who reported a lack of social support at work were 2.4 times more likely to die within those 20 years.

It's impossible to say where the relationship starts (perhaps unhealthy people stay in toxic situations longer than healthy ones). *But,* if you needed a compelling reason to quit a job full of nasty nudniks, here it is.[10]

Get a Purple Heart

Vets who received a purple heart in World War II or the Korean War live longer than those who did not receive a citation, according to research by the Department of Veterans Affairs. VA researchers studied 10,255 vets in an effort to learn what makes some more resilient to posttraumatic stress syndrome and early death than others. The results were surprising. "Among the older veterans we studied, those with the Purple Heart citations had half the mortality rate of those without Purple Heart citations," explained lead author Dr. Tim Kimbrell in a VA press release.[11] Does bravery (or bravado) lead to longevity?

Live in a city

Americans who live in urban areas have a slightly higher life expectancy than citizens who live in rural areas: 77.8 years versus 76.8 years, according to research by the *Wall Street Journal*.[12]

Get a degree

Here's another reason to get a college degree: you'll likely live longer. Education is a big determinant of longevity, according to one of longest-running longitudinal studies of health, run by Dr. George Vaillant of Harvard Medical School. Vaillant's study looked at the lives of 237 Harvard graduates and 332 inner-city Bostonians from 1937 to 1964. The study found that education trumps money when it comes to health. "Despite great differences in parental social class, college-tested intelligence, current income and job status, the health decline of the 25 inner-city men who obtained a college education was no more rapid than of the Harvard College graduates," said Vaillant in a Harvard press release on the study.[13]

The Healthiest Man in America

In 2011, *Men's Health* magazine searched for the healthiest man in America and selected Eric Foxman, of Madison, Wisconsin. Mr. Foxman is a 5 foot, 7 inch Jewish, married, white guy. Since health is associated with a long life, let's see which attributes made Mr. Foxman the magazine's winner:

He's Jewish. According to *Men's Health,* a 2010 study in the *Journal of Health and Social Behavior* found that 40 percent of people who were raised Jewish said their health was "excellent." Also, the *Journal for the Scientific Study of Religion* said Jews have a life expectancy of about 83 years—longer than Catholics, Protestants, and atheists.

He's married. Several studies have shown that married men live longer.

He's a nonsmoker. Smoking can reduce life expectancy by nearly 9 years, according to a 2007 study published in *Tobacco Control.*

He's a righty. The *Journal of Epidemiology and Community Health* found that this trait was associated with a longer life among athletic adult men, while *Clinical EEG & Neuroscience Journal* found a higher incidence of cardiac "abnormalities" among lefties.

He's white. Caucasians are less likely to suffer from obesity and obesity-related diseases, according to research from the Centers for Disease Control. White males live almost 5.5 years longer then black males, according to a 2010 National Vital Statistics Report.

He's under 6 feet. NBA take note: Men over 6 feet tall have twice as many blood clots as men standing below 5 foot 8, according to a study in the *American Journal of Epidemiology*. Shorter, smaller bodies have lower death rates and fewer diet-related chronic diseases, especially past middle age, according to a report in *Life Science* journal.

He has a modest waist! People with a waist circumference greater than 40.7" generate 85 percent more inpatient charges than those with a waist circumference less than 32.7", according to a study published in *Obesity Research*. Foxman's belt size is 31 inches.

Your time is limited, so don't waste it living someone else's life. Don't be trapped by dogma—which is living with the results of other people's thinking. Don't let the noise of others' opinions drown out your own inner voice. And most important, have the courage to follow your heart and intuition. They somehow already know what you truly want to become. Everything else is secondary.
—Steve Jobs, Stanford University commencement address, 2005

To lengthen your life span, DON'T:

Be obese

Obesity leads to serious health conditions such as diabetes, heart disease, and certain types of cancer, which are bound to cut a life short. More than one-third third of American adults and 17 percent of children are obese.[14] According to a 2011 report from the National Research Council, obesity might be partly to blame for the life expectancy gap between the United States and other countries (remember, the United States ranks 50 in life expectancy).

Eat red meat

Red meat doesn't just clog your arteries, it can actually increase your risk of premature death, according to a study published in the *Archives of Internal Medicine*. The study found that each daily serving of red meat (beef, pork, lamb, or a processed meat, such as bacon, bologna, hot dog, salami, or sausage) increased the risk of premature death by about 12 percent. Processed meats like hot dogs and bacon were the most deadly. The authors noted: "Compared with red meat, other dietary components, such as fish, poultry, nuts, legumes, low-fat dairy products, and whole grains, were associated with lower risk."[15]

Live in a noisy place

Living with persistent, loud environmental noise can shave months, even years off your life, according to the World Health Organization (WHO). Excessive noise contributes to all sorts of maladies, including cardiovascular disease, cognitive impairment in children, sleep disturbances, and tinnitus. Yikes! In Western Europe, 1 million healthy life years are lost every year from traffic-related noise, according to WHO.[16]

Smoke

Lifelong smoking lowers your life expectancy by more than 4.3 years after the age of 50 for men, 4.1 years for women. The average loss of life from one cigarette: 11 minutes.

Be homeless

The stats are sobering. Being homeless shaves 22 years off the life expectancy of men, and 17 of women, according to a homeless register of 32,711 people in Denmark. The study found that the death rate for homeless women was 6.7 times higher than for the general population; for homeless men, it was 5.6 times higher.[17]

Watch excessive amounts of TV

Watching the telly for an average of 6 hours a day could shorten your life expectancy by almost 5 years, according to a study of Australian adults published in the *British Journal of Sports Medicine.* Researchers used data from the Australian Bureau of Statistics and the Australian Diabetes, Obesity, and Lifestyle study and found that every hour spent watching TV lowered life expectancy by nearly 22 minutes for those 25 and older. Could sitting in front of a computer all day have similar effects? Time and a good research study will tell.[18]

I don't want to achieve immortality through my work.
I want to achieve it through not dying.
—Woody Allen

Time Trivia: Walking speed and longevity

The pace at which an older person walks may predict how long they live. Researchers examined how quickly adults over age 65 walked and found that those who clipped along at 2.25 mph (1 meter per second) or faster, lived longer than adults of the same age and sex who walked more slowly. The results of the study, which were published in the *Journal of the American Medical Association* in 2011, were based on an analysis of nine studies that looked at the walking speed, sex, age, body mass, and survival rate of nearly 34,500 older adults.

Book Excerpt: A key to survival

Although it's important to live in the moment, it's also valuable to have a vision of a promising future—particularly when the present is not that pleasant. Psychiatrist Viktor Frankl spent 3 years in Nazi prison camps during World War II and wrote about his experiences in his book *Man's Search for Meaning*. When Frankl felt desperate and despondent, he would fantasize about his life after the camps. In one vision he saw himself standing on the platform of a well-lit stage giving lectures on the psychology of the concentration camp. The ability to see beyond the present was, he believed, crucial to his survival. "The prisoner who had lost faith in the future—his future—was doomed. With his loss of belief in the future, he also lost his spiritual hold; he let himself decline and became subject to mental and physical decay." Frankl notes: "It is a peculiarity of man that he can only live by looking to the future—*sub specie aeternitatis*. And this is his salvation in the most difficult moments of his existence, although he sometimes has to force his mind to the task."[19]

Dying is something we human beings do continuously, not just at the end of our physical lives on this earth.
—Elisabeth Kübler-Ross

Survival rules

How much can an adult endure? If you're ever stranded in the wilderness, are caught in a burning building, or find that your scuba tank has run out of oxygen, remember these rules courtesy of *National Geographic* magazine:[20]

1. Humans can survive for just 2 to 3 minutes without air, but with training it's possible to hold your breath for 11 minutes.

2. Humans can survive for just 10 minutes at 300°F (children can only survive a few minutes at 120°F).

3. Humans can endure barely 30 minutes of exposure to 40°F water.

4. Humans can survive for up to 7 days without water.

5. Humans can survive for about 45 days without food.

Time Trivia: Defying Survival Rules

Prahlad Jani, an Indian holy man, claims he has not had a bite to eat since 1940 (Jani was born in 1929). In 2010, Jani was brought to a hospital in India and watched over by a team of medics for 15 days. The medics said that the yogi did not eat, drink, or go to the bathroom for the entire stay. Jani was also observed in 2003 and the findings were the same.

The Life Spans of Animals

How long do animals live?

 Pop Quiz: Match the animal to its life span

Animal	Life span in the wild
Crocodile, American	100+ years
Beluga whale	80 years
Tarantula	Up to 70 years
Zebra	Up to 60 years
Beaver	Up to 70 years

Spotted salamander	Up to 45 years
Honeybee	35 to 50 years
Octopus	Up to 30 years
Black widow spider	28 years
Elephant, Asian	25 years
Bluebird	24 years
Wolf	Up to 20 years
Ant	15 years
Elephant, African	6 to 10 years
Electric eel	6 to 8 years
Chimpanzee	Up to 5 years
Bald eagle	1 to 3 years
Tortoise, Galapagos	1 to 2 years
Parrot	3 weeks, to 3 years

Answers:

Galapagos tortoise, 100+ years; parrot, 80 years; African elephant, up to 70 years; Asian elephant, up to 60 years; American crocodile, up to 70 years; chimpanzee, up to 45 years; beluga whale, 35 to 50 years; tarantula, up to 30 years; bald eagle, 28 years; zebra, 25 years; beaver, 24 years; spotted salamander, up to 20 years; electric eel, 15 years; bluebird, 6 to 10 years; wolf, 6 to 8 years; honeybee, up to 5 years; black widow spider, 1 to 3 years; octopus, 1 to 2 years; ant, 3 weeks to 3 years

I have done my share. It is time to go. I will do it elegantly.
—Albert Einstein, explaining to doctors that
he did not wish surgery to repair his damaged aorta.
He died four days later on April 18, 1955, at 76.

Dog Years

The old axiom was simple: 1 human year equals 7 dog years. The new axiom is a bit more nuanced. Experts now believe dogs mature quickly and reach adulthood in the first two years of life—so the first year of a dog's life is more like 15 in a human's life. Then the rate slows. What's more, aging depends on a dog's breed and size; small breeds mature at a slower rate than large breeds. Here's a handy chart that does the math for you.[21]

Translating Dog Years into Human Years

Age of dog	Small breed age in human years	Medium breed age in human years	Large breed age in human years
1	15	15	15
2	24	24	24
3	28	28	28
4	32	32	32
5	36	36	36
6	40	42	45
7	44	47	50
8	48	51	55
9	52	56	61
10	56	60	66

Time Trivia: When was time divided into A.D. and B.C.?

A monk by the name of Dionysus Exiguus introduced a new calendar system based on the birth of Jesus Christ around A.D. 532. Up until this time, calendar eras were based on how long a consul had been in office and were often named for that person. Exiguus designated the years after the birth of Christ as Anno Domini, A.D., *the year of our Lord.* The years before Christ was born were called *Before Christ,* or B.C.

Why bother? Exiguus wanted to expunge the memory of Diocletian, a Roman emperor who persecuted Christians, and whose name was attached to an era. Specifically, Exiguus wanted to replace an old Easter table that had been created in the so-called Diocletian era. The last year of the old table, Diocletian 247, was immediately followed by the first year of his table, A.D. 532.

The Anno Domini era did not catch on until well after Exiguus's death. The first person to use the A.D. and B.C. nomenclature was believed to be the Venerable Bede, a monk, scholar, and author of *The Ecclesiastical History of the English People.* Bede began using the Christian calendar in his writing around the eighth century, about 1,300 years ago.

The Rewards of Time

Aging is not a total downer. Three experts explain how the mind and spirit improve over the years.

Dr. Marc Agronin is a geriatric psychiatrist and author of *How We Age: A Doctor's Journey into the Heart of Growing Old:*

"We think of aging as loss and degradation, yet we are growing and changing constantly. There is always the potential to shape your destiny. Most people experience a greater sense of well-being as they age, they make better decisions, they experience less stress, their emotions are more stable. People become less ideological in the way they think, more open to new ideas. All of this gives older individuals a wisdom that younger ones don't have. There's more to learn from the fountain of age, than from the fountain of youth."

Dr. Oury Monchi is a neurologist at the Geriatric Institute at the University of Montreal. Dr. Monchi recently published a study that examined how the brain adapts to changing situations and compared the language abilities of a group of young adults (18 to 35) to a group of older ones (55 to 75):[22]

"The older brain has experience and knows that nothing is gained by jumping the gun . . . When it comes to certain tasks, the brains of older adults can achieve very close to the same performance as those of younger ones. We now have neurobiological evidence showing that with age comes wisdom and that as the brain gets older, it learns to better allocate its resources. Overall, our study shows that Aesop's fable about the tortoise and the hare was on the money: being able to run fast does not always win the race—you have to know how to best use your abilities. This adage is a defining characteristic of aging."[23]

Barbara Strauch is author of *The Secret Life of the Grown-Up Brain:* "The brain, as it traverses middle age, gets better at recognizing the central idea, the big picture. If kept in good shape, the brain can continue to build pathways that help its owner recognize patterns and, as a consequence, see significance and even solutions much faster than a young person can."[24]

Keep in mind how little time there is, how little time there always was. Then try to be brave. Try to be someone else. Someone better.

—Will Eno, from his play *Thom Pain (based on nothing)*

Test Your Time Knowledge

Whether you read *The Book of Times* word for word, skimmed it quickly, or are just looking at it for the first time (!), relax, have fun, and enjoy the quiz. Take it alone or with a friend or best of all with a crowd of your favorite smart alecks.

If you get one question wrong, or can't answer a single one correctly, don't fret. *A man's errors are his portals of discovery* (so spoke James Joyce).

1. **How much time do women spend getting dressed on Monday? On Friday?**

   a.) 76 minutes; 19 minutes

   b.) 30 minutes; 30 minutes

   c.) 30 minutes; 10 minutes

   d.) 13 minutes; 7 minutes

2. **How long does a hug last?**

   a.) 1.5 seconds

   b.) 33 seconds

   c.) 3 seconds

   d.) too long

3. **What's the average wait time to see a doctor?**

   a.) 24 minutes

   b.) 36 minutes

   c.) 11 minutes

   d.) 1 hour and 10 seconds

4. **By what percentage do annoyed customers overestimate their waiting time?**

   a.) 101 percent

   b.) 12 percent

   c.) an unreasonable amount

   d.) 23 to 50 percent

5. **What is the happiest time of the day?**

   a.) Cocktail hour

   b.) First thing in the morning

   c). Quitting time

   d.) Early morning and near bedtime

6. **How long does the first rush of romantic love last?**

   a.) 1 month

   b.) About 2 years

   c.) 63 days

   d.) 1.5 years

7. **How long does foreplay last on average? Sexual intercourse?**

   a.) 16.9 minutes; 19.2 minutes

   b.) too long; not long enough

   c.) 13 minutes; 12 minutes

   d.) 6 minutes; 21 minutes

8. **At what age are kids cutest?**

   a.) 3.5 years old

   b.) 4.5 years old

   c.) 3 months old

   d.) 1 year old

9. **How long does a mattress last?**

   a.) 7 to 10 years

   b.) 5 to 6 years

   c.) It depends on the type of sleeper you are

   d.) 15 years

10. **How much time does a washing machine save?**

    a.) 2 hours a week

    b.) 212 hours a year

    c.) none

    d.) 2.5 hours a week for singles, 3.5 for families

11. **When should you toss out your sunscreen?**

    a.) After 3 months

    b.) When the bottle is empty

    c.) When it gets gloppy

    d.) After 1 year

12. **How long would it take for termites to destroy a wood home?**

    a.) 10 to 15 years, but significant damage could occur in 3 years

    b.) 1 year

    c.) 20 years, but damage could occur within a month

    d.) depends on where you live

13. **How long would it take to build a log cabin from a kit (a real log cabin, not a toy cabin)?**

    a.) 3 days

    b.) 9 weeks

    c.) 12 weeks

    d.) 6 months

14. **For how many years was Sony's Walkman produced?**

    a.) 12 years

    b.) 25 years

    c.) 31 years

    d.) It's still being made

15. **How long does it take for your fingernail to grow an inch?**

    a.) 10 months

    b.) 2 months

    c.) 30 days

    d.) 68 days

16. **How many skin cells fall off your body in one minute?**

    a.) 1 million

    b.) 1,200

    c.) 3,222

    d.) 50,000

17. **How long does it take to digest a cocktail?**

    a.) 24 hours

    b.) 1 hour

    c.) 20 minutes

    d.) 12 minutes

18. **How long did Bonnie and Clyde's crime spree last?**

    a.) 21 months

    b.) 3 years

    c.) 12 weeks

    d.) 3.5 years

19. **What's the worst month to be in the hospital?**

    a.) July

    b.) January

    c.) December

    d.) August

20. **How much time does our mind spend wandering?**

    a.) 12.2 percent of the day

    b.) 2 hours a day

    c.) 33.3 percent of the day

    d.) 46.9 percent of our waking hours.

21. **What's the highest number of hours an employee is allowed to work each day, according to federal law?**

    a.) 8 hours

    b.) 12 hours, but with overtime pay after 8

    c.) There is no limit on daily work hours.

    d.) 7 hours

22. **How often are workers interrupted?**

    a.) Every 3 minutes

    b.) Five times a day

    c.) Every 20 minutes

    d.) Once an hour

23. **When is the best time to schedule a meeting?**

    a.) Monday at 10 A.M

    b.) Tuesday at 3 P.M.

    c.) At lunchtime

    d.) First thing in the morning

24. **How long did it take Leonardo da Vinci to paint *The Last Supper*? *The Mona Lisa*?**

    a.) 12 years, 2 years

    b.) 10 years, 1 year

    c.) 12 months, 7 months

    d.) 3 years and 20 years

25. **How long did it take to build the Empire State Building?**

    a.) 1 year and 45 days

    b.) 3 years, 2 months

    c.) 6 years, 12 days

    d.) 2 years

26. **How long did it take for the *Titanic* to sink?**

    a.) 12 hours

    b.) 3 hours and 10 minutes

    c.) 35 minutes

    d.) 2 hours and 40 minutes

27. What's the best age at which to make financial decisions?

a.) 53 years old

b.) 33 years old

c.) 40 years old

d.) 45 years old

28. Which country has the highest life expectancy?

a.) Japan at 88 years

b.) Monaco at 90 years

c.) United States at 78 years

d.) Japan and Switzerland are tied in first place

29. How much do you reduce your life by smoking one cigarette?

a.) By a year

b.) By 3 months

c.) By 11 minutes

d.) By 1 hours, and 10 minutes

30. How quickly do most teens expect a reply to a text?

a.) 10 seconds

b.) Within an hour

c.) Within 5 minutes

d.) 12 minutes

Answers:

1. a
2. c
3. a
4. d
5. d
6. b
7. a
8. b
9. a
10. c
11. d
12. a
13. b
14. c
15. a
16. d
17. b
18. a
19. a
20. d
21. c
22. a
23. b
24. d
25. a
26. d
27. a
28. b
29. c
30. c

Acknowledgments

First and foremost, I want to thank the many wise experts I called upon while writing this book. Their brilliant research and willingness to share their findings made *The Book of Times* possible. I'm grateful to Marc Agronin, Paul Amato, Roy F. Baumeister, Gary Belsky, Suzanne Bianchi, Rabbi Yaacov Deyo, Robin Dunbar, K. Anders Ericsson, Terri D. Fisher, William Frey, Kate Griffiths, Sandra Hofferth, Bill Kosteas, Alan Lax, Jonathan Levav, Robert Levine, Gloria Mark, Kara McCarthy, David Meyer, Cassie Mogilner, Donald A. Norman, Paul Reber, Josiah D. Rich, M.D., John Robinson, Mark Showalter.

In addition, these professionals were exceptionally generous with their time and expertise: Kathleen DeBoer, Alana K. Johnson, Marc Mauer, Amy Sapola, M.D., Shmuel Shoham, M.D., Tom Snyder, Ted Socha, and double thanks to Mark Weaving and Micha Godard, both of whom endured numerous pesky emails from me for nearly two years.

Thanks to my lovely and astute agent, Jennifer Gates, for believing in this project from the get-go, to Stephanie Meyers for securing the book's future, and to Jessica McGrady for helping to shape the book into a cohesive tome. A special thanks to Cole Hager, who arrived like a knight in shining armor during the book's final phase; he brought all the pieces (and people) together and to a happy conclusion.

Jessie Edwards, publicist extraordinaire, brought the book out to the world; thanks for your diligence and hard work.

Kudos to Kristina DiMatteo and Mumtaz Mustafa, who created the playful cover design.

Thanks to Joan Caplin for her clever contributions to the Creativity chapter, to Amanda Angel for unearthing arcane sports facts and to Claudia Bloom for her sleuth work on the Home chapter.

A bear hug to Will McGreal for his unwavering confidence and sage psychic support; he always knew this was possible.

My wonderful husband gave me the time to write and the support I needed to finish the book. (Not to mention letting me take over his home office!) I love you for who you are and for who you have helped me become.

Notes

Chapter : Introduction

1. Jacques Derrida, *Points . . .: Interviews, 1974–1994* (Palo Alto, CA: Stanford University Press, 1995), p. 116.

Chapter 1: Daily Life

1. Bureau of Labor Statistics (BLS), American Time Use Survey (ATUS), 2011.
2. Mark Aguiar and Erik Hurst, "Measuring Trends in Leisure: The Allocation of Time Over Five Decades," National Bureau of Economic Research Working Paper No. 12082, 2006.
3. L. Copeland, "Americans Give Thumbs Up to Free Time," *USA Today,* December 23, 2010.
4. Almudena Sevilla, et al., "Leisure Inequality in the United States: 1965–2003," *Demography*, Vol 49, No 3, 2012, p 939.
5. http://www.ropercenter.uconn.edu/data_access/tag/leisure_and_recreation.html.

6. K. Fisher and J. Robinson, "Daily Routines in 22 Countries: Diary Evidence of Average Daily Time Spent in Thirty Activities" (Oxford, UK: Centre for Time Use Research, University of Oxford, 2010).

7. Jayoti Das and Stephen DeLoach, "Mirror, Mirror on the Wall: The Effect of Time Spent Grooming on Earnings." Electronic copy available at http://ssrn.com/abstract=1013649.

8. Richard Wrangham, *Catching Fire: How Cooking Made Us Human* (London: Profile Books, 2009), p. 141.

9. Survey by the British department store Debenhams, 2010.

10. *The Daily Mail,* August 15, 2011, http://www.dailymail.co.uk/femail/article-2025905/Women-spend-hours-EVERY-DAY-gossiping-claims-study.html.

11. Vanessa Wight, Joseph Price, Suzanne Bianchi, and Bijou Hunt, "The Time Use of Teenagers," *Social Science Research* 38, no. 4 (June 2009): 792–809.

12. BLS, ATUS, 2010. Summary Table 2.

13. BLS, ATUS, 2003–2011.

14. Cheryl Russell, New Strategist Publications, April 17, 2012, e-mail.

15. Gallup Daily Tracking, June 2010; Gallup, February 2012.

16. Pew Forum's 2007 U.S. Religious Landscape Survey.

17. University of Chicago National Opinion Research Center's General Social Survey (GSS), 2004.

18. BLS, ATUS, averages for 2005–2009.

19. General Social Survey, 2008.

20. National Endowment for the Arts, "Beyond Attendance: A Multi-Modal Understanding of Arts Participation," May 2007–May 2008.

21. 2008 MSPA Wait Time Study, Mystery Shopping Providers Association, North America.
22. Press Ganey, *2010 Medical Practice Pulse Report,* p. 15.
23. Press Ganey, *2010 Pulse Report Emergency Departments,* p. 6.
24. Tom Krause, "The Waiting Game," Maritz Research, 2008.
25. Carl Bialik, "Justice—Wait for it—on the Checkout Line," *The Wall Street Journal,* August 19, 2009. In a 1988 study at a downtown Boston bank, Prof. Larson and colleagues showed that customers overestimated their wait times by 23 percent. When researchers added a clock displaying expected wait time, customers' estimates of wait time were more accurate. A 1997 study of two Sacramento supermarkets coauthored by Gail Tom showed that at one of them, customers overestimated waiting time by over 50 percent.
26. Robin Dunbar, "You've Got to Have 150 Friends," *The New York Times,* December 25, 2010.
27. Robin Dunbar, *How Many Friend Does One Person Need?* (Harvard University Press, November 2010).
28. Gallup-Healthways Well-Being Index.
29. Scott Golder and Michael Macy, "Diurnal and Seasonal Mood Vary with Work, Sleep, and Daylength Across Diverse Cultures," *Science,* Vol. 333 no. 6051 (2011) pp. 1878–1881.
30. http://press.friendsreunited.co.uk/33magicnumber.
31. Robert Levine, *A Geography of Time: On Tempo, Culture, and the Pace of Life* (New York: Basic Books, 1998).

Chapter 2: Love

1. Enzo Emanuele, et al. "Raised Plasma Nerve Growth Factor Levels Associated with Early-Stage Romantic Love," *Psychoneuroendocrinology*, Volume 31, Issue 3. (2006), 288–294.

2. D. Marazziti and D. Canale, "Hormonal changes when falling in love," *Psychoneuroendocrinology*, 29 (2004), 931–6.

3. Michael Eisenberg et al.,"Socioeconomic, Anthropomorphic, and Demographic Predictors of Adult Sexual Activity in the United States: Data from the National Survey of Family Growth," *Journal of Sexual Medicine,* 2010.

4. T. Fisher, Z. Moore, and M. Pittenger, "Sex on the Brain?: An Examination of Frequency of Sexual Cognitions as a Function of Gender, Erotophilia, and Social Desirability," *Journal of Sex Research,* 2011.

5. Emese Nagy, "Sharing the moment: The Duration of Embraces in Humans," *Journal of Ethology*, 29, 2 (2011) 389–393.

6. Durex – *Global Sexual Wellbeing Survey*, 2011; Weiss and S. Brody, "Women's partnered orgasm consistency is associated with greater duration of penile-vaginal intercourse but not of foreplay," *Journal of Sexual Medicine* 6 (2009) 135–41; "Sex and the American Man," *Esquire,* April 2012.

7. E. W. Corty and J. M. Guardiani, "Canadian and American Sex Therapists' Perceptions of Normal and Abnormal ejaculatory latancies: How Long Should Intercourse Last?," *Journal of Sexual Medicine,* 5 (5), (2008), 1251–6.

8. Tomothy Ferriss, *The 4-Hour Body: An Uncommon Guide to Rapid Fat-Loss, Incredible Sex, and Becoming Superhuman* (Crown Archetype, 2010), p. 237

9. Roberto Refinetti, "Time for Sex: Nycthemeral Distribution of Human Sexual Behavior." *Journal of Circadian Rhythms* 3, no. 1 (2005): 4.

10. Survey was conducted by online sex toy retailer, Lovehoney.co.uk, in 2012.

11. Helen Fisher et al., "Reward, Addiction, and Emotion Regulation Systems Associated with Rejection in Love." *Journal of Neurophysiology* 104 (2010): 51–60.

12. National Health Statistics Reports, Number 49, March 22, 2012.

13. U.S. Census Bureau, Current Population Survey, March, and Annual Social and Economic Supplements, 2010 and earlier; U.S. Census Bureau, 2009 American Community Survey.

14. 2011 Real Weddings Survey, TheKnot.com.

15. "Wedding Statistics Everyone Should Know," EngagementRingQuestions.com; November 18, 2008.

16. Paul Amato, *Alone Together: How Marriage in America Is Changing* (Cambridge, MA: Harvard Universty Press, 2007).

17. From "Forget the Seven-Year-Itch, The Breaking Point for Couples Comes After THREE Years," *Daily Mail Online,* March 8, 2011.

18. Esure survey, reported in "Had a Row with Your Partner Today? That'll Be One of the 2,455 You Will Have This Year," *The Daily Mail,* May 20, 2011.

19. Dietrich Klusmann, "Sexual Motivation and the Duration of Partnership," *Archives of Sexual Behavior* 31, no. 3 (2002): 275–87.

20. U.S. Census Bureau, "Marital Events of Americans: 2009," August 2011.
21. U.S. Census Bureau, "Number, Timing, and Duration of Marriages and Divorces: 2009."
22. National Health Statistics Reports, Number 49, March 22, 2012, p. 7.
23. John Tierney, "Refining the Formula That Predicts Celebrity Marriages' Doom," *The New York Times,* March 12, 2012.
24. Organisation for Economic Co-operation and Development, "Mean Duration of Marriage to Divorce, 1960 to 2008," Chart SF3.1.F.
25. U.S. Census Bureau, "Number, Timing, and Duration of Marriages and Divorces: 2009," Table 8.
26. Pew Research Center, "Barely Half of U.S. Adults Are Married—A Record Low," December 14, 2011.

Chapter 3: Family

1. Suzanne M. Bianchi, "Family Change and Time Allocation in American Families," Alfred P. Sloan Foundation, Focus on Workplace Flexibility, November 2010.
2. U.S. Census Bureau, "A Child's Day: 2009."
3. William Doherty, "Share the Table: The State of Dinner Time in America" (a white paper study commissioned by Barilla), 2008, p. 27.
4. Bianchi, p. 11.
5. Families and Work Institute, "2008 National Study of the Changing Workforce."

6. Gretchen Livingston and Kim Parker, "A Tale of Two Fathers," Pew Research Center, June 15, 2011.

7. Meredith Phillips, "Parenting, Time Use, and Disparities in Academic Outcomes," *Whither Opportunity,* Russell Sage Foundation, September 2011.

8. Li Zhu Luo, "Are Children's Faces Really More Appealing Than Those of Adults? Testing the Baby Schema Hypothesis Beyond Infancy," *Journal of Experimental Child Psychology* 110, no. 1 (September 2011): 115–24.

9. National Research Council, "Prevent Reading Difficulties in Young Children," 1998.

10. Scholastic, *2010 Kids & Family Reading Report.*

11. "Fewer Mothers Prefer Full-Time Work," Pew Social Trends, July 12, 2007.

12. Bureau of Labor Statistics, ATUS 2005-09, Table A-7.

13. Daniel Kahneman et al., "A Survey Method for Characterizing Daily Life Experience: The Day Reconstruction Method," *Science* 306 (2004): 1776–80.

14. Kerstin Aumann et al., "The New Male Mystique," Families and Work Institute, 2011.

15. Ibid.

16. Diane Swanbrow, "Exactly How Much Housework Does a Husband Create?," *Michigan Today,* April 3, 2008.

17. eSure survey, January 7, 2010, published in *The Daily Mail:* "The 40-Minute Chore Wars."

18. Shira Offer and Barbara Schneider, "Revisiting the Gender Gap in Time-Use Patterns: Multitasking and Well-Being among Mothers and Fathers in Dual-Earner Families," *American Sociological Review* Vol 76, No 6 (2011), p. 809–833.

19. Katherine Laughon et al., "Induction of Labor in a Contemporary Obstetric Cohort," *American Journal of Obstetrics and Gynecology*, 2012.

20. Daniel Sacks and Betsey Stevenson, "What Explains the Rapid Increase in Childcare Time: A Discussion of Gary Ramey and Valerie A. Ramey's 'The Rug Rat Race,'" *Brookings Papers on Economic Activity*, 2010.

21. Email correspondense with CDC.

22. Constance T. Gager and Scott T. Yabiku, "Who Has the Time?" *Journal of Family Issues* 31, no. 2 (2010).

23. ATUS, 2011. Tables 13, 15, and 16.

24. Xinran Lehto et al., "Vacation and Family Functioning," *Annals of Tourism Research* 36, no. 3 (2009): 459–79.

25. Trip Advisor Survey, 2010.

26. http://www.summerlearning.org/.

27. John Steinbeck, *Travels with Charley* (New York: Penguin, 1962), p. 127.

28. Jeroen Nawjin et al., "Vacationers Happier, but Most Not Happier After Holiday," *Applied Research in Quality of Life*, 2010.

Chapter 4: Home

1. Coldwell Banker Real Estate Survey, May 15, 2012.

2. National Association of Realtors Home Buyer and Seller Survey, November 5, 2010.

3. U.S. Census Bureau, Survey of Income and Program Participation (SIPP), 1993, 1996, 2001, and 2004 Panels, Wave 2 Migration Topical Module.

4. Jessica Lautz from the National Association of Realtors crunched these numbers.

5. Association of Home Appliance Manufacturers; National Association of Home Builders/Bank of America Home Equity Study of Life Expectancy of Home Components.

6. Ibid.

7. National Association of Home Builders/Bank of America Home Equity Study of Life Expectancy of Home Components.

8. International Sleep Products Association.

9. National Association of Home Builders/Bank of America Home Equity Study of Life Expectancy of Home Components

10. BugInfo.com; Alan Lax at the USDA's Agricultural Research Service

11. "History of Sears Modern Home," SearsArchives.com.

12. From *Room for Debate,* "Books You Can Live Without," *The New York Times,* December 27, 2009.

13. Source: Stilltasty.com.

Chapter 5: Body

1. "Modern Bodies," *New Scientist,* March 21, 2011.

2. Axel Kramer et al., "How Long Do Nosocomial Pathogens Persist on Inanimate Surfaces? A Systematic Review," *BMC Infectious Diseases,* August 16, 2006.

3. W. C. D. Maile and K. J. L. Scott, "Digestibility of Common Foodstuffs As Determined by Radiography," *The Lancet,* January 5, 1935, and June 29, 1935.

4. www.calorieking.com.

5. Research by Ad Vingerhoets, professor of clinical psychology at Tilburg University in the Netherlands and William H. Frey, a neuroscientist and biochemist professor, Department of Pharmaceutics, University of Minnesota. Cited in Katherine Rosman, "Read It and Weep, Crybabies," *Wall Street Journal,* May 4, 2011.

6. National Sleep Foundation; American Academy of Sleep Medicine.

7. http://www.npr.org/2011/05/16/136275658/late-to-bed-early-to-rise-makes-a-teen-sleepy.

8. CDC, "Short Sleep Duration Among Workers—United States, 2010," *Morbidity and Mortality Weekly Report* 61, no. 16 (April 27, 2012): 281–85.

9. "Animal Sleep," *National Geographic,* September 2011.

10. http://faculty.washington.edu/chudler/chasleep.html.

11. http://www.wnlc.com/mindbender.htm.

12. http://well.blogs.nytimes.com/2011/10/03/really-the-claim-yawning-cools-the-brain/.

13. Paul Taylor, "Nap Time," Pew Research Center, July 29, 2009.

14. Amber Brooks and Leon Lack, "A Brief Afternoon Nap Following Nocturnal Sleep Restriction: Which Nap Duration is Most Recuperative?" *SLEEP*, Vol 29, No. 6 (2009).

15. Medscape's 2012 Physician Compensation Report.

16. National Hospital Discharge Survey, *Health 2010,* from the U.S. Department of Health and Human Services, Table 102, p. 337.

17. Eric A. Weiss, M.D. et al., "Drive Through Medicine," *Annals of Emergency Medicine,* 2009.

18. John Q. Young, M.D., et al. "July Effect: Impact of the Academic Year-End Changeover on Patient Outcomes: A Systematic Review," *Annals of Internal Medicine*, Vol 155, No. 5 (2011).

19. D. Morra et al., "U.S. Physician Practices Versus Canadians: Spending Nearly Four Times As Much Money Interacting with Payers," *Health Affairs* 30, no. 8 (August 2011): 1443–50.

20. Interview with Dr. Jay Giedd for *Frontline* series, "Inside the Teenage Brain," PBS.org.

21. Health United States, 2010, Table 57.

22. Matthew Killingsworth and Daniel Gilbert, "A Wandering Mind Is an Unhappy Mind," *Science,* Vol 330, No. 6006 (2010), p. 932

23. Britta K. Holzel, et al, "Mindfulness Practice Leads to Increases in Regional Brain Gray Matter Density," *Psychiatry Research: Neuroimaging*, 191 (2011), p. 36.

24. Shai Danziger, Jonathan Levav, and Liora Avnaim-Pesso, "Extraneous Factors in Judicial Decisions," *Proceedings of the National Academy of Sciences,* May 1, 2011.

Chapter 6: Occupation

1. *Digest of Education Statistics,* 2011; Table 8.

2. *Digest of Education Statistics,* 2011: Introduction.

3. U.S. Dept. of Education, "Average Length of School Day . . . ," National Center for Education Statistics (NCES), Schools and Staffing Survey (SASS), Public School Questionnaire, 2007–2008.

4. NCES, "Historical Summary of Public Elementary and Secondary School Statistics: Selected Years, 1869–70 through 2007–08."

5. International Association for the Evaluation of Educational Achievement (IEA), "Trends in International Mathematics and Science Study" (TIMSS), 2007.

6. U.S. Department of Education NCES, "Schools and Staffing Survey 2007–08."

7. The OECD Programme for International Student Assessment (PISA), 2009.

8. TIMSS, 2007, IEA.

9. *Digest of Education Statistics,* 2010, Table 415.

10. "The World of Seven Billion," *National Geographic,* March 2011.

11. Homework numbers are from Sandra Hofferth, PhD, director of Maryland Population Research Center.

12. Philip Babcock and Mindy Marks, "Leisure College, USA: The Decline in Student Study Time," American Enterprise Institute for Public Policy Research, No. 7, August 2010.

13. Pew Research Center, *Is College Worth It?,* Pew Research Center Publications, May 15, 2011.

14. Ibid.

15. Bureau of Labor Statistics, American Time Use Survey, High School Students and Homework.

16. NCES, *The Baccalaureate and Beyond Longitudinal Study.*

17. *The College Completion Agenda: 2011 Progress Report,* published by the CollegeBoard.

18. "Pathways to Prosperity," Harvard Graduate School of Education, February 2011.

19. "Trends in College Pricing 2011," CollegeBoard.

20. *Time Is the Enemy,* 2011 national report from Complete College America, p. 12.
21. Bureau of Labor Statistics, Employment Status of the Civilian Noninstitutional Population, Household Data Annual Averages.
22. "The Working Day: Understand Work Across the Life Course," www.bc.edu/agingandwork.
23. Mark Ellwood, "Hours Worked by Job," *Get More Done Time Study Consulting,* Pace Productivity.
24. Joan C. Williams and Heather Boushey, "The Three Faces of Work-Family Conflict," Center for American Progress, January 2010.
25. Organisation for Economic Co-Operation and Development (OECD), "Average Annual Working Time," Table 8, September 26, 2012.
26. OECD, Better Life Index, Work-Life Balance, data is from 2012 or latest available year.
27. Mika Kivimäki et al., "Using Additional Information on Working Hours to Predict Coronary Heart Disease: A Cohort Study," *Annals of Internal Medicine* 154 (April 5, 2011): 457–63.
28. Bureau of Labor Statistics, Employee Tenure, January 2012.
29. http://www.gallup.com/poll/154178/expected-retirement-age.aspx.
30. Retirement Survey, Wells Fargo & Company, released November 16, 2011.
31. R. Morin, "The Impact of Long-Term Unemployment," July 22, 2010, Pew Research Center, July 22, 2010.
32. comScore Media Metrix.
33. "Wasting Time at Work 2008," Salary.com.

34. Ibid.
35. Victor Gonzalez and Gloria Mark, " 'Constant, Constant, Multi-tasking Craziness': Managing Multiple Working Spheres," *Proceedings of ACM CHI'04*, Vienna, Austria, April 26–29, 2004, pp. 113–20; and Gloria Mark, Victor Gonzalez, and Justin Harris,"No Task Left Behind? Examining the Nature of Fragmented Work," *Proceedings of ACM CHI'05*, Portland, OR, April 2–7, 2004, pp. 321–30.
36. http://whenisgood.net, October 20, 2009.
37. *Dave Barry Turns 50,* p. 183.
38. Mercer Consulting, "Employee Statutory and Public Holiday Entitlements—Global Comparisons," London, October 13, 2009.
39. "Taking a Break," Ipsos Global @dvisory; August 6, 2010.
40. Monster Global Poll, October 2010.
41. U.S. Census Bureau, "Mean Travel Time to Work of Workers 16 Years and Over Who Did Not Work at Home," 2011 American Community Survey, estimates.
42. INRIX National Traffic Scorecard Report 2010.
43. The 20 cities in the survey were Bangalore, Beijing, Buenos Aires, Chicago, Johannesburg, London, Los Angeles, Madrid, Mexico City, Milan, Montreal, Moscow, Nairobi, New Delhi, New York City, Paris, Shenzhen, Singapore, Stockholm, and Toronto.
44. Steve Crabtree, "Wellbeing Lower Among Workers with Long Commutes," Gallup-Healthways Well-Being Index, August 13, 2010.

Chapter 7: Creation

1. Translated by John Frederick Nims
2. www.pbs.org/treasuresoftheworld/guernica/glevel_1/ gtimeline.html.
3. Jeffrey K. Smith and Lisa F. Smith, "Spending Time on Art," *Empirical Studies of the Arts* 19, no. 2 (2001): 229–36.
4. National Endowment for the Arts (NEA), "Time and Money: Using Federal Data to Measure the Value of Performing Arts Activities," April 2011.
5. Ibid.
6. Ibid., p. 19.
7. "Miss Mitchell, 49, Dead of Injuries," *The New York Times*, August 17, 1949.
8. copyright.gov/help/faq/faq-duration.html#duration.
9. Richard Kostelanetz, *Conversing with John Cage* (New York: Routledge, 2003).
10. 2006 Music USA: National Association of Music Merchants (NAMM) Global Report.
11. "2009 Public Attitudes Towards Music," commissioned by the NAMM and conducted by Gallup.
12. http://www.wannaplaymusic.com/gallup_poll.
13. http://www.slashfilm.com/by-the-numbers-the-length-of-feature-films/.
14. http://www.nationmaster.com/encyclopedia/Buracz-Bosnitz.
15. http://www.cinemetrics.lv/database.php.
16. http://www.rockyhorror.com/history/timeline.php.
17. Ryan Devlin, "Let's Do the Time Warp Again. And Again. And Again," *Premiere* 18, no. 9 (June 2005): 58–60, 62–63.

18. newyorktransportation.com.
19. Susan Stamberg, "The Golden Gate Bridge's Accidental Color," National Public Radio, April 26, 2011
20. Katherine Rosman, "The Most Awkward Meeting," *Wall Street Journal,* May 19, 2011.

Chapter 8: Energy

1. Richard Schulz and Christine Curnow, "Peak Performance and Age Among Superathletes: Track and Field, Swimming, Baseball, Tennis, and Golf," *Journal of Gerontology* 43, no. 5 (1988): P113–P120. Baseball: J. C. Bradbury, "Peak Athletic Performance and Ageing: Evidence from Baseball," *Journal of Sports Science* 27, no. 6 (2009): 599–610. Tennis: The Sony Ericsson WTA Tour 10-year age eligibility and professional development review, http://www.ncbi.nlm.nih.gov/pmc/articles/PMC2653876/. Tennis men: http://bleacherreport.com/articles/499816-roger-federer-is-he-in-his-prime-what-are-a-tennis-superstars-best-years-1. Football: http://thesportdigest.com/archive/article/when-does-football-player-get-old. Basketball: http://www.basketballprospectus.com/unfiltered/?p=399. Hockey: http://www.arcticicehockey.com/2010/1/21/1261318/nhl-points-per-game-peak-age. Soccer: http://www.robystahl.com/page/Reaching-the-Peak.aspx.
2. Tennis: "Going to College or Turning Pro? Making an Informed Decision!," USTA. Basketball: "Exit Discrimination in the NBA: A Duration Analysis of

Career Length." Baseball: http://www.sciencedaily.com/releases/2007/07/070709131254.htm. Hockey: http://www.quanthockey.com/Distributions/CareerLengthSeasons.php. Football: http://thebiglead.com/index.php/2011/04/22/nfl-career-length-and-average-age-versus-average-life-expectancy/.

3. http://www.time.com/time/asia/2006/heroes/ae_khan.html.

4. http://icc-cricket.yahoo.net/rules_and_regulations.php.

5. NEA report, *Time and Money,* April 2011.

6. David Biderman, "11 Minutes of Action," *The Wall Street Journal*, January 15, 2010.

7. C. D. Mah et al., "The Effects of Sleep Extension on the Athletic Performance of Collegiate Basketball Players," *Sleep* 34, no. 7 (2011): 943–50.

8. "Surfing into Jaws," National Geographic Adventure, July 2002.

9. GuinessWorldRecords.com.

10. http://www.environmentalgraffiti.com/featured/swimming-in-the-coldest-waters-on-earth/6238.

11. http://www.towerrunning.com/english/races/newyork.htm.

12. Doug Reed in a Nike blog post, March 30, 2011.

13. CDC, "Leisure-Time Physical Activity," March 2012.

14. M. J. Reifschneider, K. S. Hamrick, and J. N. Lacey, "Exercise, Eating Patterns, and Obesity: Evidence from the ATUS and Its Eating & Health Module" *Social Indicators Research* 101 (2011): 215–19.

15. http://www.cdc.gov/physicalactivity/everyone/guidelines/adults.html.

16. S. J. Marshall, "Translating Physical Activity Recommendations into a Pedometer-Based Step Goal," *American Journal of Preventive Medicine* 36, no 5 (May 2009): 410–15.

17. http://www.cdc.gov/healthyyouth/physicalactivity/facts .htm.

18. Alpa V. Patel et al. "Leisure Time Spent Sitting in Relation to Total Mortality in a Prospective Cohort of U.S. Adults," *American Journal of Epidemiology,* Vol 172, Issue 4, p. 419.

19. H.D. Sesso et al, "Physical Activity and Coronary Heart Disease in Men: The Harvard Alumni Health Study," *Circulation,* 102, 9 (2000), p. 975–80.

20. http://www.healthstatus.com/calculators.html.

Chapter 9: Destruction

1. *Sourcebook of Criminal Justice Statistics Online,* http://www .albany.edu/sourcebook/pdf/t6132010.pdf.

2. Bureau of Justice Statistics, Felony Sentences in State Courts, 2006.

3. Paul Guerino, Paige M. Harrison, and William J. Sabol, *BJS Statisticians,* "Prisoners in 2010," Table 5.

4. John Bronsteen et al., "Happiness & Punishment," *University of Chicago Law Review* 76 (2009).

5. I. A. Binswanger et al.,"Release from Prison—A High Risk of Death for Former Inmates," *NEJM* 356 (2007): 157–65.

6. R. Reinhold, "The Longest Trial—A Post Mortem; Collapse of Child-Abuse Case: So Much Agony for So Little," *The New York Times,* January 24, 1990.

7. U.S. Department of Justice, "Temporal Patterns of Bank Robbery, FBI Bank Crime Statistics, 2010."

8. Jane Gross, "Ferdinand Marcos, Ousted Leader of Philippines, Dies at 72 in Exile," *The New York Times,* September 29, 1989.

Chapter 10: Money

1. Federal Reserve; United States Mint.

2. http://www.bankrate.com/calculators/retirement/401-k-retirement-calculator.aspx.

3. Harvard Graduate School of Education, *Pathways to Prosperity,* footnote 6: These figures are taken from "Education Pays 2010," which was prepared by the College Board Advocacy and Policy Center and released in September 2010.

4. Harvard Graduate School of Education, *Pathways to Prosperity*, February 2011.

5. http://www.businessweek.com/articles/2012-04-19/does-it-pay-to-study-at-state.

6. http://www.businessweek.com/interactive_reports/mba_roi_2011.html.

7. Cassie Mogilner, "The Pursuit of Happiness: Time, Money and Social Connection," *Psychological Science* 21, no. 9 (September 2010): 1348–54.

8. Jayoti Das and Stephen B. DeLoach, "Mirror, Mirror on the Wall: The Effect of Time Spent Grooming on Earnings," June 8, 2009. Electronic copy available at: http://ssrn.com/abstract=1013649.

9. Vasilios D. Kosteas, "The Effect of Exercise on Earnings: Evidence from the NLSY," *Journal of Labor Research* (2012) 33:225–250

10. Sumit Agarwal, John C. Driscoll, Xavier Gabaix, and David Laibson, "The Age of Reason: Financial Decisions over the Life-Cycle with Implications for Regulation," prepared for *Brookings Papers on Economic Activity,* October 19, 2009.

11. http://www.nber.org/cycles/cyclesmain.html.

12. http://www.gallup.com/poll/153995/Stay-Home-Moms-Lean-Independent-Lower-Income.aspx.

13. http://www.salary.com/what-s-a-mom-worth-in-2012/.

14. Jennifer Aaker, Melanie Rudd, and Cassie Mogilner, "If Money Doesn't Make You Happy, Consider Time," *Journal of Consumer Psychology* 21, no. 2 (April 2011): 126–30.

Chapter 11: Media

1. Retrevo online survey, March 2010.

2. Pew Research Center's Project for Excellence in Journalism, The State of the News Media, 2011.

3. CBS/*Vanity Fair* poll, October 2010.

4. "Average hours spent per visitor on social networking sites across geographies," comScore Data Mine, April 2011.

5. Pew Research Center's Internet & American Life Project.

6. Nielsen, NPOWER, Live+7, Prime (9/21/11–1/29/12).

7. http://blog.nielsen.com/nielsenwire/media_entertainment/what-time-is-really-primetime/.

8. Nielsen, NPOWER, Live+7, Prime (9/21/11–1/29/12).

9. NPD *Gamer Segmentation 2010.*

10. Entertainment Software Association.

11. "The State of the News Media 2011," Pew Research Center's Project for Excellence in Journalism.

12. "Americans Spending More Time Following the News," Pew Research Center for People & the Press, September 12, 2010.

13. Heather Chaet, "The Tween Machine," *AdAge*, June 25, 2012.

14. Virginia Rideout et al. *Generation M2: Media in the Lives of 8- to 18-Year-Olds* (Menlo Park, CA: Henry J. Kaiser Family Foundation, 2010).

15. Ibid.

16. Kaiser Family Foundation.

17. Pew Research Center, "Teens, Smartphones and Texting," March 19, 2012.

18. AT&T survey as part of the "It Can Wait Campaign," May 14, 2012.

19. Ibid.

20. Power Reviews, *The 2011 Social Shopping Study.*

21. Happycustomer, a STELLA Service publication, June 27, 2012.

22. American Express, "2012 Global Customer Service Barometer."

23. http://researchnews.osu.edu/archive/multitask.htm.

24. Wilhelm Hofmann, Roy Baumeister, Georg Förster, and Kathleen Vohs, "Everyday Temptations: An Experience Sampling Study of Desire, Conflict, and Self-Control in Everyday Life," *Journal of Personality and Social Psychology,* published online December 12 , 2011.

25. Janna Anderson and Lee Rainie, "Millennials will benefit and suffer due to their hyperconnected lives," Pew Internet & American Life Project, February 29, 2012.

Chapter 12: Longevity

1. *CIA World Fact Book,* 2012 estimate.
2. American Human Development Project of the Social Science Research Council; data from 2010–11.
3. http://www.socialsecurity.gov/planners/lifeexpectancy .htm.
4. http://gosset.wharton.upenn.edu/mortality/second_ version.html.
5. L. Wilhelmsen et al., "Factors Associated with Reaching 90 Years of Age: A Study of Men Born in 1913 in Gothenburg, Sweden," *Journal of Internal Medicine,* 2010.
6. A. Trichopoulou and E. Vasilopoulou, "Mediterranean Diet and Longevity," *British Journal of Nutrition* 84, suppl. 2 (2000): S205–S209.
7. The study was conducted by the National Institutes of Health and American Association of Retired People (AARP).
8. Chi Pang Wen et al., "Minimum Amount of Physical Activity for Reduced Mortality and Extended Life Expectancy: A Prospective Cohort Study," *The Lancet,* October 2011.
9. H. S. Friedman and L. R. Martin, *The Longevity Project* (New York: Hudson Street Press, 2011).

10. A. Shirom, "Work-Based Predictors of Mortality," *Health Psychology,* May 2011.
11. Tim Kimbrell et al., "The Impact of Purple Heart Commendation and PTSD on Mortality Rates in Older Veterans," *Depression and Anxiety* 28, no. 12 (December 2011): 1086–90.
12. Melinda Beck, "City vs. Country: Who Is Healthier?," *The Wall Street Journal,* July 12, 2011
13. http://news.harvard.edu/gazette/2001/06.07/01-happywell.html.
14. http://www.cdc.gov/obesity/data/trends.html.
15. An Pan et al., "Red Meat Consumption and Mortality," *Archives of Internal Medicine,* 2012.
16. World Health Oganization, "Burden of Disease from Environmental Noise," 2011.
17. John R. Geddes and Seena Fazel, "Extreme Health Inequalities: Mortality in Homeless People," *The Lancet,* June 14, 2011.
18. J. Lennert Veerman et al., "Television Viewing Time and Reduced Life Expectancy: A Life Table Analysis," *British Journal of Sports Medicine,* August 2011.
19. Viktor E Frankl, *Man's Search for Meaning* (New York: Pocket Books, 1984).
20. S. Sperry, "Human Limits," *National Geographic* magazine, 2009.
21. Purina, "Your Dog's Age in Human Years."

22. Ruben Martins, France Simard, Jean-Sebastien Provost, and Oury Monchi, "Changes in Regional and Temporal Patterns of Activity Associated with Aging during the Performance of a Lexical Set-Shifting Task," *Cerebral Cortex,* August 24, 2011.

23. "Clinical study shows young brains lack the wisdom of their elders," press release from University of Montreal, August 25, 2011.

24. Barbara Strauch, "How to Train the Aging Brain," *The New York Times,* December 29, 2009.